FEELING MEDIATED

INSTITU

CRITICAL CULTURAL COMMUNICATION

GENERAL EDITORS: SARAH BANET-WEISER AND KENT A. ONO

Feeling Mediated

A History of Media Technology and Emotion in America

Brenton J. Malin

NEW YORK UNIVERSITY PRESS

New York and London

NEW YORK UNIVERSITY PRESS
New York and London
www.nyupress.org

References to Internet websites (URLs) were accurate at the time of writing.
Neither the author nor New York University Press is responsible for URLs that
may have expired or changed since the manuscript was prepared.

Library of Congress Cataloging-in-Publication Data

Malin, Brenton J., 1972-
Feeling mediated : a history of media technology and emotion in America / Brenton J.
Malin.
pages cm. -- (Critical cultural communication)
Includes bibliographical references and index.
ISBN 978-0-8147-6279-0 (cloth : alk. paper) -- ISBN 978-0-8147-6057-4 (pbk. : alk. paper)
1. Communication and technology--United States--History. 2. Mass media and technology-
-United States--History. 3. Communication--Psychological aspects--United States. 4. Mass
media--Psychological aspects--United States. 5. Mass media and culture--United States. I.
Title.
P96.T422U6358 2014
  302.23--dc23
                          2013042413

New York University Press books are printed on acid-free paper,
and their binding materials are chosen for strength and durability.
We strive to use environmentally responsible suppliers and materials
to the greatest extent possible in publishing our books.

Manufactured in the United States of America

10 9 8 7 6 5 4 3 2 1

Also available as an ebook

CONTENTS

ACKNOWLEDGMENTS

Despite the great amount of feelings we project onto our technologies, most of our emotional support still comes from other people. I could have written this book without a computer, as impossible as the hype of the twenty-first-century "digital age" might make that seem. But I could never have done so without the support provided to me by a great number of students, teachers, colleagues, friends, and family members— overlapping categories, to be sure. At New York University Press, Eric Zinner, Alicia Nadkarni, and Ciara McLaughlin—who was instrumental in helping me develop this manuscript in its early and later stages— offered important editorial direction and guidance, as did the series editors, Kent Ono and Sarah Banet-Weiser, and Dorothea Halliday was enormously supportive and helpful through the production process. Bruce Gronbeck, John Peters, Paddy Scannell, Michael Vicaro, and Jonathan Sterne—who offered me some of my most sustained guidance on the manuscript as a whole—all read and commented on portions of the manuscript at various stages of development, helping me hone and develop my arguments throughout.

My colleagues in the Communication Department at the University of Pittsburgh have supported this project in a variety of ways, most notably by creating the engaging intellectual climate that helped me think through the various questions I take up in the following pages. I thank Lynn Clarke, Donald Egolf, William Fusfield, Olga Kuchinskaya, John Lyne, Gordon Mitchell, Lester Olson, John Poulakos, Shanara Reid-Brinkley, Barbara Warnick, and Ronald Zboray. Other colleagues have offered a range of additional forms of support and guidance, engaging me in conversation, asking me invaluable questions, suggesting additional readings, and otherwise allowing me to more deeply engage my subject matter over the past several years. I thank Mark Lynn Anderson, Misha Antonich, James Batcho, Mazviita Chirimuuta, Gordon Coonfield, Hugh Curnutt, Daniel Emery, Jane Feuer, Lucy Fischer, Michael

Glass, Joshua Gunn, Chul Heo, Robert Hellyer, David Henry, Alexandra Hui, Robert MacDougall, Daniel Morgan, David Nye, David Park, Mark Paterson, Jefferson Pooley, Richard Popp, Peter Putnis, Adam Roth, Eric Rothenbuhler, Joseph Sery, Peter Simonson, and Lance Strate. Likewise, my undergraduate and graduate students at the University of Iowa, St. Olaf College, Allegheny College, San Francisco State University, and the University of Pittsburgh have consistently helped me to think through and develop these and other ideas, providing me with great everyday examples of technology use and offering their own interesting theories of our emotional connections to the media. If not for the insights and energy I draw from my students in the classroom, I would never have been able to complete this book.

Portions of this book appeared in earlier form as "Looking White and Middle-Class: Stereoscopic Imagery and Technology in the Early Twentieth-Century United States," *Quarterly Journal of Speech* 93, no. 4 (2007): 403–24; "Electrifying Speeches: Emotional Control and the Technological Aesthetic of the Voice in the Early 20th Century U.S.," *Journal of Social History* 45, no. 1 (2011): 1–19; and "Mediating Emotion: Technology, Social Science, and Emotion in the Payne Fund Motion Picture Studies," *Technology and Culture* 50, no. 2 (2009): 366–90. Likewise, many of the ideas from these chapters were presented at conferences of the National Communication Association, the International Communication Association, and the Media Ecology Association. All of these occasions provided me with invaluable feedback that has greatly informed the arguments and analyses expressed here.

Finally, I thank those who give me emotional support every day: my wife, Jessie, and my sons, Mason and Graham. You are my constant reminder that our most meaningful emotional connections happen through face-to-face encounters with other people.

Introduction

If punctuation can capture the spirit of a time, then none has done so as clearly for the digital age as the emoticon. The idea that users communicating through high-tech screens would need a hieroglyphic to represent their moods or facial expressions suggests a series of tensions at work in our digital connections to each other. According to Marvin Minsky, author of *The Emotion Machine: Commonsense Thinking, Artificial Intelligence, and the Future of the Human Mind,* "it is still widely believed that minds are made of ingredients that can only exist in living things, that no machine could feel or think, worry about what might happen to it, or even be conscious that it exists—or could ever develop the kinds of ideas that could lead to great paintings or symphonies."[1] A computer scientist, Minsky hopes to challenge this popular dogma by demonstrating that human emotions function like complex machine operations and, in turn, that machines could be capable of their own complex emotions. If Minsky is correct about the current public attitude toward machines, emoticons could be explained as an attempt on the part of digital communicators to inject human emotion into an apparatus they ultimately view as incapable of expressing it. Interacting with and through what they apparently view as an emotionless screen, users try their best to re-create a human face.

At the same time, the widespread use of emoticons might imply a stronger public faith in technological expressions of emotion than Minsky's comments suggest. A recent study illustrates that some users may be more comfortable interacting with less realistic avatars, and that in certain cases they prefer the emoticon to the actual human face. The study had students interact in several different settings: in a highly realistic videoconferencing setting, in an audio-only setting, and in a setting using an unrealistic "emotibox" that rendered a user's facial expressions as abstract computer graphics. In the end, those users who had interacted through the emotibox not only disclosed more, but perceived their partners as more "revealing, honest, and friendly" than those who interacted through the video screen.[2] Far from cold, emotionless machines, these users seemed to experience the expressions of the avatars with which they interacted as genuine emotions. In *The Media Equation: How People Treat Computers, Television, and New Media Like Real People and Places*, Byron Reeves and Clifford Nass note a number of similar studies that evidence people's willingness to engage machines emotionally, including smiling back at grinning avatars and being more interactive with computers they perceive as polite.[3] For the users in these studies, technological representations of emotion were not simply degraded forms of actual human feeling. They were their own unique versions of emotion that mimicked, and in certain instances surpassed, more apparently human emotional expressions.

This book explores a range of assumptions about the capacity of communication technologies to capture, convey, and express emotion. A culture's communication technologies have a complex relationship with how its people understand and talk about their own and others' feelings. Although Minsky suggests that emotion has generally been seen as a uniquely human attribute, at odds with technology, just as often machines have been seen as *better carriers of human feeling*— as people's responses to "friendly computers" might indicate. These complicated understandings of emotion parallel equally complicated notions about technology itself. At the same time that new technologies are imagined to be improving human life by increasing people's capacity for memory, movement, and so forth, they are also regularly seen to be hurting humanity by overloading people with information and generally bombarding them with new—and presumably dangerous— stimulation. There is something both wonderful and creepy about the mechanical emotions Minsky identifies, and it is this tension—and specific attempts to resolve it—that centers the discussion in the following chapters.

In addition to the complicated machine emotions implied in the emoticon, U.S. popular culture abounds with stories of feeling machines. In Stanley Kubrick's 1968 film *2001: A Space Odyssey*, the computer HAL 9000 displays a highly developed emotionality that roots much of the narrative tension of the film. In one of the film's most dramatic scenes, the astronaut David Bowman disconnects HAL's cognitive circuits after HAL has killed his fellow astronaut, Frank Poole. As Dave begins the process of shutting "him" down, HAL comments in an eerily calm but desperate voice, "I know everything hasn't been quite right with me, but I can assure you now, very confidently, that it's going to be all right again. I feel much better now. I really do." As Dave continues, HAL's pleading persists: "Will you stop, Dave? Stop, Dave. I'm afraid. I'm afraid, Dave."

In this scene, as throughout much of the film, HAL seems the most emotionally engaged of the crew. Despite HAL's pleadings, Dave continues his work without a word, mechanically turning screw after screw until HAL is shut down. In a last expression of emotion, HAL sings "Daisy Bell," a late nineteenth-century love song, as his voice gradually gives out. Illustrating his own understanding of machine emotions, Kubrick explained shortly after *2001* was released that he had aimed to depict "the reality of a world populated—as ours soon will be—by machine entities who have as much, or more intelligence as human beings, and who have the same emotional potentialities in their personalities as human beings."[4] HAL may indeed be the most human character of the film in terms of his overt emotional expression.

Kubrick's collaboration with Steven Spielberg, *A.I.* (2001), offers a similar story of machines that emulate—and in many ways surpass—human emotional expression. In a future in which highly developed robots serve humans in various ways, the electronics company Cybertronics develops a robotic young boy capable of feeling and expressing love. A prototype named David is adopted by a family whose son, Martin, is in a long-term coma. Although she is initially skeptical about this apparent replacement son, Martin's mother, Monica, eventually warms to David, and goes through the sequence of words programmed to activate his imprinting process. Once she does, David, who has called her Monica to this point in the film, calls her "Mommy," and his intense love begins.

Like HAL, David is among the most emotionally expressive characters in his film world. The other boys, including Martin, who eventually
returns home as David's brother, are depicted as hypertypical preado-
lescent sociopaths. Martin convinces a naïve David to cut off a lock of
their mother's hair while she sleeps, promising that she will love him
more if he does. One of Martin's friends stabs David's hand to see if
he is equipped with pain sensors, prompting David to grab Martin and
cower behind him until they both fall into a swimming pool, nearly
drowning. But if David is the most sensitive of these boys, his robotic
teddy bear, Teddy, is often all the more so. In one scene, at Martin's
prompting, David and Martin both call to Teddy, attempting to prove
which of them Teddy likes more. Exasperated, Teddy escapes when
Monica walks through the room. "Mommy," he cries, as he grabs her
hand. The more mechanical the character, the film seems to suggest, the
more emotionally sensitive they are.

Numerous other stories have depicted similarly emotional machines.
Despite the initial portrayal of Arnold Schwarzenegger's cold, calculating robotic killer in *Terminator* (1984), by the film's sequel (*Terminator 2:
Judgment Day*, 1991), this time-travelling cyborg has turned to good and
found his sensitive side. "I now know why you cry," the distraught Terminator tells John Connor, the boy it returns from the future to protect, just
before it destroys itself to keep its technology from falling into the wrong
hands. Computers and robots are not the only machines imagined to feel
and express emotions. Wilbert Awdry's *Railway Series* books, which date
back to the 1940s and include the popular children's character Thomas
the Tank Engine, depict railroad cars and other vehicles that feel joyful,
sad, grumpy, and a range of other emotions. The Disney film *The Love
Bug* (1968), and subsequent films in the same series, depict a highly emotional, loving automobile named Herbie, whose horn, headlights, and
windshield wipers become his tools of emotional expression.

Playing on this tendency to attribute emotions to machines, an IKEA
television commercial from the early twenty-first century tells the story
of an old desk lamp discarded in favor of a new model. Slow, sad music
accompanies a set of shots in which the old lamp, left on the curb outside in the rain, seems to "stare" into the house of its former owner,
where the newer, shinier lamp is in place. The punch line comes at the
end of the commercial, when a narrator walks on screen to address

the audience. "Many of you feel bad for this lamp," he begins. "That is because you're crazy. It has no feelings, and the new one is much better." Frightening, funny, cute, and eerie—popular culture has presented a wide range of feeling machines.

What do these complex beliefs regarding emotional machines, apparently held by computer scientists, film directors, and the audience for IKEA lamp commercials, tell us about attitudes regarding both emotion and technology? What are the roots of these desires and fears that technologies might respond to us with emotion? Although HAL and David are only the products of science fiction, the average twenty-first-century American interacts with and through a wide range of technologies. Automated voices "greet" us on our telephones. Computerized GPS units can guide us as we drive. Even our friendships can become computerized versions of themselves. When we talk on most cell phones, we hear a digital approximation of our friends' and family members' voices. In many ways, an army of HALs and Davids may seem to be mediating the whole range of our emotional connections.

As with the emoticon, the extent to which we believe a technology capable of capturing and expressing human emotions will bear upon the quality of our interactions with each other. If I believe that an iPhone provides some unique expression of my individualized tastes and feelings, how might I alter my communications with others or even my sense of my own emotional life? In contrast, if I see these new technologies as offering degraded or even malevolent forms of emotional connection, how might I behave in order to cope with these presumably lessened emotional interactions? In short, how do I approach the world of communication, technology, and emotion if I view an iPhone as Herbie, HAL, David, Teddy, Thomas, an IKEA lamp, or something else entirely?

In the United States, for the greater part of the twentieth and early twenty-first centuries, media technologies have been given a great responsibility for human emotion. Phonograph records, movies, radio broadcasts, television programs, websites such as Facebook, and a range of other media have been both celebrated and criticized for their power to communicate emotion. A range of clergy, teachers, politicians, and others have held these media accountable for everything from teen suicides to mass murder. The more powerful or advanced the technology, these thinkers often claim, the more dangerous its emotional stimulations. At the same time, the idea of the global village, whether attributed to Marshall McLuhan, Al Gore, Apple, or a host of others who have made similar arguments, rests on assumptions about the ability of communication technology to transcend time and space and draw people into a community of shared affections. In this view, better access to more powerful technologies seems a path toward a happier, more united world. In these, as in many other cases, both technological naysayers and celebrants seem to maintain the connection between more powerful technologies and more powerful expressions of emotion, with more advanced technologies getting us closer to our emotional hell, or heaven, respectively.

These discussions of the emotional power of communication technology often focus on the "new technologies" of a given era—those developing media that seem to be transforming a culture's abilities to connect in ways that can only be imagined. As I will illustrate in the next chapter, when the telegraph first emerged, a number of American commentators suggested that it would unite the world in one common heart, creating the sort of global village McLuhan would discuss in the following century. Others worried that it would destroy local communities and neighborhoods and create a hyperactive culture of information addicts, predicting a series of criticisms that would accompany the rise of the Internet. Each of these effects no doubt took place in certain contexts and to various extents; however, both sides would have been hard pressed to prove that the extreme transformations they predicted had arrived.

These nineteenth-century discussions, like the twentieth- and twenty-first-century ones that followed them, are part of a still longer history of Western cultural attitudes toward technology and emotion. Socrates lived at a moment when writing was a relatively new communication technology and Greece was transitioning from a predominantly oral to a written culture.[5] The complexities of this intermediary period are borne out in Plato's dialogues—*written* pieces in which Socrates, Plato's mentor, often attacks the effects of writing on his orally based philosophical dialectic. A central element of Plato's *Phaedrus* is a written text by the speechwriter Lysias that young Phaedrus obtains and then performs for Socrates. Significantly, as Jacques Derrida has noted,[6] Socrates uses the Greek term *pharmakon* (Φάρμακον) in an early reference to the alluring power of Lysias's written speech.[7] In using this word, which translates to drug, medicine, or charm—it is the original source for the English word pharmacy—Socrates suggests the emotional power carried by a written text. In fact, Socrates may be the original advocate of the so-called "hypodermic needle model" of communication, which imagines communication technologies to carry an overwhelming emotional force that, like a drug, has an immediate impact on their audiences.[8] Despite their ability to deconstruct it, Lysias's script evokes a "frenzied enthusiasm" in both Phaedrus and Socrates.[9]

In Socrates's discussion, the seductive power of Lysias's speech emanates from the problematic nature of writing itself. At the beginning of the dialogue, when Phaedrus tries to deliver his own version of the speech, Socrates insists that Phaedrus read from the script he has hidden under his cloak. "Much as I love you," he tells Phaedrus, "I am not altogether inclined to let you practice your oratory on me when *Lysias himself* is here present."[10] At a later moment, Socrates instructs Phaedrus to read part of the script out loud so that he "can listen to the author himself."[11] Living at the moment that he did, Socrates would have witnessed the transition from a culture of face-to-face, present-bound communication, to one that seemed to defy the logic of time and space by allowing a speechwriter such as Lysias to be both present and absent at the same time.

Reflecting a cultural uneasiness with this transition, Socrates later suggests that one of the problems with writing is that it "doesn't know how to address the right people, and not to address the wrong."[12] This is because the author is not there to make decisions about when and with whom to communicate. In contrast, Socrates tells Phaedrus, there is a type of discourse "that knows to whom it should speak and to whom it should say nothing." Phaedrus understands this to be not the "dead discourse" of writing, "but the living speech, the original to which the written discourse may be fairly called a kind of image." In its transformation of time and space, Socrates suggested, the written word was simultaneously living and dead. It detached the emotions of language from the body of the speaker and represented them in the disembodied form of the text. Like a drug, the written word stimulated emotion without a clear source, lending it an apparently evocative and eerie power.

A number of later thinkers shared Socrates's sense of the eerie power of developing communication technologies, even if they disagreed with his more negative evaluation of it. Elocution—discussed in more detail in chapters 1 and 3—was an intricate form of oral performance that developed well after the transition to literacy, and its practitioners tended to take a more positive view not only of writing, but of printing technology as well. Elocutionists created elaborate forms of gesture and speech believed to convey highly emotional meanings to their audiences. In his 1846 *Manual of Elocution*, Merritt Caldwell suggested the centrality of printing to elocutionary practice:

> The art of engraving was not understood by the ancients. In modern works on elocution much advantage has been taken of the improvements of this art; and in regard to gesture, abundant illustrations have been furnished, which addressing the eye, make a stronger as well as a more definite impression on the mind than could well be made by words.[13]

In contrast to Socrates's evaluation of writing, for Caldwell the new technologies of printing and engraving had enhanced rather than degraded oral communication. Printing was a technological condition that made possible the emotional power of elocutionary speech.

Despite this apparent disagreement, Socrates and the elocutionists shared an essentially pharmacological understanding of communication technology. For Socrates, the written form of Lysias's speech gave it an unusual power over the emotions. For Caldwell and his fellow elocutionists, the orderliness of print, not to mention the gestures and movements those printed pages captured, carried its own emotional power. Both Socrates and the elocutionists largely took for granted that newer communication technologies brought about an increased stimulation of emotion, Socrates largely reacting against it, and the elocutionists largely embracing it. For both, there was something special about the new technologies of their age that seemed to herald a new emotional climate.

Why might someone assume that a newer communication technology would cause more powerful emotional stimulation? Both the positive and negative views on the telegraph mentioned above shared this belief with Socrates and the elocutionists, and many contemporary views on the Internet and other "new media" do so as well, as do many of the other thinkers discussed in the chapters that follow. Such beliefs in the interconnection of technology and emotion might seem contradictory. In many ways, emotion and technology may seem completely at odds with each other. Especially in those moments when scientific and technological progress has been given a central place in society, technologies of various sorts have been seen as highly rational, objective, and calculating. In contrast, emotions have often been seen as uniquely personal, subjective, and irrational.

However, a belief in technological progress may itself lead someone to attribute an emotional power to various technologies. This faith underlies much of the horror in science fiction representations of feeling machines. Once a highly advanced, rational computer like HAL develops a fear of dying, there appears to be little that can stop him. He can put the rational power of his technological thinking to the service of his private, irrational impulses. In a similar way, David of *A.I.* seems to love with a fervor that his human counterparts can only imagine. Stripped of human frailties, he becomes a singular loving machine. Similarly, people's belief in a technology's ability to store information, transmit messages, or make connections across time and space may suggest its ability to stimulate emotion for either good or ill. In these ways, a faith in scientific progress and the powerful rationality of new technologies can encourage concerns—or celebrations—regarding frenzied enthusiasm.

It is tempting to assume that one's contemporary moment is an apex of technological development. With no knowledge of the many technologies that would follow, Socrates may well have seen writing as an end point in the history of communication. The elocutionists seemed to associate a level of cultural and communicative perfection with the printing press, as if the ideal form of communication had finally arrived. Looking back on these earlier moments, we of the twenty-first century may assume that Socrates's head would simply explode if he were faced with the Internet, cell phones, or any number of other digital technologies, as if in our time *we really have reached a summit of communication possibilities*. For better or worse, many have suggested, today we connect with each other and exchange information at a peak rate of speed, giving our contemporary culture a heightened sense of emotional intensity.

To be sure, some thinkers assume that newer communication technologies *disconnect* people. Although Socrates attributed a kind of emotional power to writing, he also seemed to worry about how it separated people from each other. Lysias's absence from Socrates's exchange with Phaedrus was a central element in that dialogue's critique of writing. Those who assumed that the telegraph would destroy communities worried that people would choose distant communications over connecting with their neighbors. The use of emoticons may evidence a concern for technological and emotional disconnection as well—as suggested by Minsky's comments above—by reflecting worries that computers replace interpersonal communication with a technology that cannot quite sustain it. Whether people believe that new technologies enhance or hinder connections, and whether they see those effects as positive or negative, will illustrate a range of assumptions about the interconnection of emotion and technology.

The chapters that follow explore some of the complexities of these views of emotion and technology, focusing on how they have impacted thinking about communication in the United States. In contrast to the majority of the sources I analyze, I do not take a position on the relative advancement of various "new technologies" or on whether these technologies enhance or hinder our connections to each other. Likewise, I do not take a position on the relative worth of "emotional" versus "rational" communication. As I will argue throughout, such positions too often camouflage larger assumptions about culture or identity. People's claims about the emotional power of a specific communication technology may have little to do with the technology itself and much more to do with concerns about the moment in which they are living.

Rather than trying to prove or refute the emotional power of any particular technology, this book aims to analyze and demonstrate the consequences of various *rhetorics* of emotion and technology. By rhetoric, I mean the ways our language—whether in scientific papers, advertisements, movies such as *A.I.*, or other means by which we communicate—shapes our views of ourselves and the world around us and create particular possibilities for being-in-the-world. At first thought, technology and emotion might seem anything but rhetorical. Technologies have clear physical, mechanical, or electrical properties that make them suitable for some tasks and not others. A knife is significantly different from a telegraph key, and the printing press is different from a telephone. A culture communicating by carrier pigeon would have different possibilities for connecting than one using the Internet. In the same way, emotions have very real physiological characteristics. We feel emotions in our bodies, often as physical experiences that seem to defy language altogether.

Neither technologies nor emotions exist in a vacuum, however. Whether a given technology is good or bad is a product of argument as much as it is its specific physical properties. It is for this reason that people can come to such disparate positions on the cultural worth of the telegraph, radio, Internet, or other communication technologies. Likewise, most of these technologies arrive surrounded by a host of messages about their value and use. It would be hard to separate the technological characteristics of the iPhone from the advertisements through which Apple shapes its meaning, as well as from its discussion in newspapers, television programs, and other media sources. Similarly, a nineteenth-century citizen would likely have used a telegraph with knowledge of the celebrations and denunciations that surrounded it at the time.

Of particular importance to these discussions is what Leo Marx has called the *rhetoric of the technological sublime* (a concept I explore in more detail in the next chapter).[14] Focusing on the nineteenth century, Marx illustrates some complex and often contradictory ways that American writers addressed the new technologies of the locomotive, telegraph, and steam ship. These technologies were seen as sublime because they seemed to dwarf both individuals and the vast American pasture. While Marx focuses his discussion on nineteenth-century America, much of the same could be said about Socrates, as well as many of the twentieth- and twenty-first-century authors I discuss. The presumed sublime power of technology has served many as a rhetorical trope through which to both celebrate and vilify the technologies of their day, allowing people to shape the rhetoric of technology in their own particular ways.

Emotional expressions are also heavily influenced by the historical moment and culture in which they take place. Despite contemporary ideas about stoic masculinity, for example, at various times in the past crying has been seen as especially masculine.[15] The historian Peter

Stearns has addressed these ideas through his concept of "emotionology," which he defines as "the attitudes or standards that a society, or a definable group within a society, maintains towards basic emotions and their appropriate expression," including the "ways that institutions reflect and encourage these attitudes in human conduct." His work recognizes how everything from parenting manuals to dating etiquette can establish dominant ideals about different emotional expressions.[16] Although these ideals cannot necessarily force someone to feel emotions in a particular way, they can set strong limitations on acceptable emotional displays, and thus create consequences for those who step outside them (for instance, the contemporary American male who weeps more than is considered masculine). Still, work in anthropology and cultural neurology suggests that even the physiological experiences of emotion can vary widely from one group to another, based on the norms and rules of each culture.[17]

Analyzing these rhetorics of emotion and technology should allow us to think more critically about how we interact with and through the communications media that surround us. Deliberations on the emotional power of technologies can have very powerful effects, not only on consumers of these technologies, but on those who produce them, as well as on the scientists, government agents, clergy, and other thinkers who attempt to make sense of them. For instance, if cultural discussions seem to focus on some specific sort of emotional stimulation, media producers may adjust their products as a way of taking advantage of it. The media producers I discuss in the following chapters responded to cultural anxieties about immigration and class with very particular marketing schemes and product adjustments.

Scientific analyses of media effects can have their own effects on a culture. If social scientists begin to believe that popular music has a negative impact on listeners' emotions, and communicate those beliefs to the larger public, then parents, educators, and legislators may well act accordingly. Likewise, in moments dominated by beliefs that better technology itself will establish stronger emotional ties between people—as was the case in many of the early twentieth-century discussions the following chapters address—communication is often conceived as a technical problem to be solved with bigger or better transmitters. Such views tend to favor the wealthiest members of society who can afford the most technologically advanced equipment. They also tend to ignore the social, ritual, and cultural elements that also make up communication practices, flattening them to a simple matter of information transmission.

Correspondingly, to the extent that discussions of emotion undertaken in scientific studies, popular magazines, advertisements, and so forth delimit appropriate levels of emotional expression, they will tend to legitimate certain kinds of communication and denigrate others. Claims labeled "too emotional," for instance, can be dismissed without consideration, as can those people who make them—an issue that has faced women at numerous points in the history of democratic and scientific debate.[18] Arguments about technologies' ability to communicate emotion reflect desires and fears about the human capacity to do the same. How a culture addresses these matters will have important consequences for their views of technology, emotion, themselves, and the world around them.

In addition to these more specific effects, discussions of emotion and technology also tell us much about the hopes and anxieties of the culture in which they take place. A culture's understanding of race, class, or gender may make it see certain groups as especially vulnerable to the emotional manipulation of different technologies. For instance, U.S. social critics and researchers have tended to be especially wary of media targeted at the working classes, women, or other presumably at-risk groups. Concerns about the nickelodeon, a cheap, early twentieth-century movie theater, reflected attitudes about motion pictures themselves as well as their predominantly working-class and immigrant audiences. The Payne Fund motion picture studies, which I explore in chapter 4, were driven in part by concerns about the vulnerabilities of working-class youth. Analyzing a culture's rhetorics of emotion and technology offers a means of understanding the complexities of meaning and identity through which its members struggle.

Given the recurrence of these discussions of emotion and technology, a whole range of moments in history could generate valuable and insightful analyses. This book focuses primarily on one interesting manifestation of these rhetorics, exploring the period of the early twentieth-century United States, from the turn of the century until the mid-1930s. The thinkers of this time shared many attitudes with Socrates, the elocutionists, and the telegraph commentators of the nineteenth century. However, several unique characteristics of the period prove especially fruitful in terms of discussing the rhetorics of technology and emotion.

For one, this period saw the birth and expansion of much of the modern electronic mass media. Thomas Edison was granted a patent for a motion picture system in 1891. Emile Berliner founded the American Gramophone Company in 1892. Guglielmo Marconi established his early American radio company, American Marconi, in 1899. Over the next several decades, motion pictures, phonographs, radio, and a range of other media technologies found wide distribution and popularity. As Friedrich Kittler has noted, the arrival of cinema and the phonograph in particular challenged a range of cultural perceptions. Unlike the writing technologies that had preceded them, phonographs and films were able to *store time* by sequencing together a collection of distinct moments. The gramophone and the cinematograph were the first technologies that could "record and reproduce the very time flow of acoustic and optical data."[19]

The idea that the new media could store and broadcast time itself prompted a range of discussions regarding the emotional intensity of the age. The sense of eerie disembodiment with which Socrates experienced a written speech was replicated by many early twentieth-century thinkers, who listened to the disembodied voices emerging from their record players or "floating through the ether" as a radio broadcast. When the American historian Lewis Mumford wrote of the connections between magic, science, and technology in his 1934 book *Technics and Civilization*, he was both addressing the long-standing history of the technological sublime and pointing to a renewed sense of mysticism that surrounded the new technologies of the period in which he was writing.[20] As had been the case for the telegraph and earlier technologies, both the utopian and dystopian bandwagons were crowded with educators, clergy, politicians, and other thinkers who made various proclamations about the effects of this new media age. For many of these commentators, theirs was a transitional moment that would either enhance or tear down the human connections they had come to know. As a result, communication technologies took a central place in many discussions of early twentieth-century culture and society, just as they have in our own.

The new media were accompanied by a like explosion in advertising. American advertising had begun an ascent in the mid- to late nineteenth century with the aggressive promotion of a variety of "patent medicines." However, the American advertising industry did not begin to take its current-day form until the 1890s, when agencies developed more specialized positions—account executives, copywriters, and so forth—and took control over more elements of the advertising process. While nineteenth-century advertisers had generally allowed the newspaper or magazine publisher who printed an ad to make decisions about images, typeface, and other design elements, the turn-of-the-century agency saw each of these as a crucial part of the advertiser's vision. As the industry grew, so did the number of advertisements in circulation. From the Civil War to the turn of the century, the revenue from advertising rose from $50 million to $500 million, and the money spent on advertising went from .7 percent of the gross national product to 3.2 percent.[21] Advertising agencies entered the twentieth century with a new sense of identity and purpose.

This growth in advertising played into the culture of "conspicuous consumption" that Thorstein Veblen identified with the American leisure class of the turn of the century.[22] The increased attention to advertising imagery—fostered both by changes in the industry and by advances in printing technology that made larger, more detailed pictures possible—encouraged stronger associations between consumer goods and various middle- and upper-class lifestyles. Soaps, colas, automobiles, radios, and other products were presented as representations of the consumer's self-identity. Consumer products were not merely goods to use, advertisers increasingly suggested; they were essential components of one's everyday identity.[23]

Closely related to the growth of advertising was the early twentieth-century formalization of another field of promotion: public relations. According to the *Oxford English Dictionary*, the phrase was first used in its current sense—to describe the general identity of an organization or important person—in 1898. By 1925, writers had explicitly recognized that "any publicity is good publicity" and that there was "no such thing as bad publicity," stressing this growing climate of promotion.[24] Edward Bernays, widely considered the father of public relations, opened a PR firm in 1919 and published his first book-length treatise on the topic in

1923. By 1935, Bernays claimed that the "organization of communication in the United States enables practically any person or any group or any movement to be brought almost immediately into the closest juxtaposition with people almost anywhere."[25]

Advertisers and other promotion experts played on this new sense of worldwide publicity. Dale Carnegie's suggestions for "how to win friends and influence people" built upon a whole range of earlier twentieth-century pronouncements regarding how people's speech, dress, personality, and other elements of self-presentation reflected their general character.[26] Early twentieth-century advertisements suggested that certain brands of automobile tires reflected a more civilized upbringing. According to Columbia, Victor and other record companies, the kind of phonograph one purchased could do the same. As publicity became increasingly important, product manufacturers seemed to insist that everyone was a walking PR stunt. In this climate, communication technologies were ways of both connecting with other people and demonstrating one's social status and high-technological sophistication. A radio both picked up broadcasts from the outside world and broadcast the social standing of its owner.

This period also saw the beginning of America's rise as a global power. Although this book deals almost exclusively with issues within the United States, the country's development as an international superpower offers an important backdrop for these domestic matters. In 1898, the Spanish-American War and the resulting Treaty of Paris left the United States with a burgeoning international empire. America's participation in World War I reiterated its importance on the world scene. The war also dealt heavy blows to the German, Russian, Ottoman, and Austro-Hungarian empires. As the United States expanded its international reach from the turn of the century into the 1910s and beyond, it enlarged its cultural influence and the market for its products. The Hollywood film industry took advantage of these new international relations, as did other media producers. Following the war, Hollywood quickly became a complex multinational enterprise, distributing its films throughout the globe.[27] The early twentieth century signaled the beginning of the global dominance of American popular culture, which gave an added impetus—and consequence—to the nation's domestic media production.

The expanding media and advertising industries were also met by a new group of American social scientists, who began to take seriously the capacity of communication technologies to capture and transmit emotion. In 1892 Yale University founded its psychological lab, which early on featured such figures as Edward Wheeler Scripture, who made recording technologies a central subject of his investigations. One of his students at Yale, Carl Seashore, who became a professor in the University of Iowa's psychological lab in 1897, devoted his career to the psychology of music. At Iowa, Seashore built a veritable cottage industry for the psychological study of communication phenomena. He and his students and colleagues analyzed phonograph records, speeches, motion pictures, vocal performances, and a range of related topics.

In addition to analyzing media technologies, Seashore and his fellow researchers also employed them in their studies. Maintaining the sublime power of technology, Seashore suggested that photographic, motion picture, and recording technologies were a psychologist's best tools for analyzing the various forms of communication studied in his lab. Using tonoscopes, phonophotography, and a range of other recording devices he and his colleagues designed, Seashore put the new communication technologies to work in analyzing themselves.[28] Two of the Payne Fund motion picture studies took place in Seashore's Iowa lab, one using motion picture–based psychological equipment to analyze the effects of motion pictures on audience members' emotions.

Scripture, Seashore, and other social scientists who took up similar topics were inaugurating what the media scholar Paul Lazarsfeld would decades later call an "administrative" approach to communication research. As Lazarsfeld defined them, administrative researchers, including himself, worked with private companies and the government to improve the marketing, publicity, and other effects of the media.[29] The sorts of early twentieth-century studies undertaken in Scripture's and Seashore's labs were both implicitly and explicitly administrative. By focusing on the power of various media to capture and transmit emotions, these studies offered implicit endorsements of the very claims media producers were marketing with their products—that specific communications media could produce specific emotional effects for their audiences. More explicitly, a number of researchers worked directly with media producers to help them improve the marketing or functioning of their goods. Seashore, in fact, found commercial success of his own, creating a widely successful test of musical talent that he sold to schools throughout the country.

These relationships between commercial media production and social scientific media research had a range of important implications for the period, as the following chapters will demonstrate. For instance, a number of media producers developed scientific or pseudoscientific explanations of the benefits of their products. Some quoted academic social researchers directly, or even got their endorsements. Those products developed by academic media researchers themselves, such as Seashore's music tests and the Pronunciphone—a set of phonograph records developed by scholars at the University of Wisconsin and intended to scientifically enhance one's pronunciation (addressed in chapter 3)—suggested still more specific connections between the new technologies and the new social science of the time. Finally, the attention of social scientists began to highlight the potential benefits of the commercial media to the education of youth and adults. Stereoscopes, motion pictures, radio programs, and phonograph records were all marketed as aids in the education process and found captive audiences in schools throughout the country.

Of course, as was the case with the Payne Fund studies, many social scientists took a negative view of the emotional impact of these new technologies. However, even these studies could reinforce the more general view put forward by the commercial media. Many of these studies still employed recording and motion picture apparatuses, suggesting the unique ability of these technologies to capture and transmit emotion. Likewise, by focusing on such issues as how commercial motion pictures created widespread emotional stimulation or deceived audiences' perceptions, these researchers reiterated the sublime power of communication technologies. These very critiques could reiterate the assumed technological power that media producers relied upon to sell their products.

These discussions of the power of communication technology impacted another significant feature of this period. From 1912 to 1934 a series of important legal and regulatory decisions took place that gave shape to America's commercial media system for the next sixty-plus years, if not longer. The most important of these, the Communications

Act of 1934, which bookends my period of analysis, was the chief legislation regulating American communication until 1996. It overwhelmingly legislated the rhetoric of technological sublimity that had been discussed by social scientists, media producers, and other concerned citizens over the previous three decades. In this act, as well as for the FCC that the act created, the "public interest" in communication was defined primarily in technological terms. A good media system was one that transmitted meanings in an effective manner.

Together, these various discussions created an interesting and in many senses troubling set of attitudes about emotion and technology. To many thinkers of the time, the new media seemed uniquely capable of both capturing and transmitting human emotions. For businesspeople who made their livings selling record players, home movie cameras, and a host of other media devices, this became a common trope through which to advertise their wares. A more sophisticated record player would presumably give its owner greater access to a whole range of sentiments they might otherwise be denied. The widespread diffusion of these media devices to homes and throughout the general culture panicked many social researchers and other critics who shared this belief in the power of technologies but were apprehensive about the emotional overstimulation they might bring about. However, even as social scientists worried about the emotional power of movies and other media, many, like Seashore and Scripture, employed motion pictures, record players, and other media devices in their research laboratories. The high-tech power that made a movie dangerous for its audiences also seemed to make motion photography ideal as an apparatus for scientifically recording emotions. In these and other ways, the new media technologies became embroiled in a complex set of celebrations and critiques.

Owing to its importance in media history, the early twentieth century has been explored by a number of media historians. Such writers as Susan Douglas, Robert McChesney, Paul Starr, Erik Barnouw, Carolyn Marvin, Daniel Czitrom, James Carey, Michele Hilmes, Lisa Gitelman, Friedrich Kittler, Jeffrey Sconce, and Jonathan Sterne have offered important investigations of this early period of American media.[30] Of these, Douglas, Carey, Marvin, Gitelman, Kittler, Sconce, and Sterne provide particularly detailed *technological* histories, making the technological features of the media they explore a central part of their discussion. By placing these media technologies in their historical context, these scholars have explored the cultural understandings of media at a time in which, as Marvin succinctly puts it, "old technologies were new." Other histories of non-media technologies, such as those by Wolfgang Schivelbusch, Leo Marx, David Nye, and Wiebe Bijker, have likewise made important strides in these directions.[31] Similarly, Jonathan Crary's studies of technology and perception in the nineteenth century have close parallels to my discussion of early twentieth-century mediated emotion.[32] While these scholars do not take emotion as a central concern, their historical investigations of media and technology have provided an important foundation for my own analyses.

At the same time, a rich body of work on the cultural history of emotion has developed over the last decade, some of which focuses specifically on this same early twentieth-century period. Although such writers as Norbert Elias and Mikhail Bakhtin had offered early histories of emotion, this work did not begin to coalesce into a coherent body of scholarship until the later part of the twentieth century. Peter Stearns's research on the history of emotion was central to this, and is an important influence on my own work. By showing how different historical moments maintained different standards regarding emotional expression, the work of Stearns and his followers offered an important means of exploring the public and social aspects of emotional experience. Because of the work of Stearns, as well as that of such scholars as Daniel Gross, Sara Ahmed, Melissa Gregg, and Gregory Seigworth, by the early twenty-first century it was possible to talk of an "affective turn" in social theory, as more and more scholars began to take the emotions seriously as a social and historical phenomenon.[33]

Feeling Mediated investigates how thinking about emotion intersects with thinking about technology, focusing primarily on an intellectual and rhetorical framework established during the early twentieth century that continues to stand in the way of larger social understandings of mediated emotion. I call this perspective *media physicalism*. The philosophical position of physicalism developed during the early twentieth century within the Vienna Circle, which included such philosophers as Moritz Schlick, Otto Neurath, and Rudolf Carnap. Taking science to be a unified language describing the reality of the world, physicalists assumed that most philosophical questions could be best answered by employing the vocabulary and methods of the natural sciences. As framed by Herbert Feigl, a student of the Vienna Circle who spent time in the University of Iowa's philosophy department during Carl Seashore's reign, physicalism understood human experience as reducible to a set of physiological traits. From the standpoint of physicalism, "to every proposition describing introspectively what, as we say, is given as a datum of my consciousness, there would be a corresponding proposition in physical language describing, as we say, the condition of my nervous system."[34] A person's conscious experiences were equivalent to a set of physical reactions in the body.

In the early twentieth-century discussions of media and emotion I address, a technologically focused brand of physicalism gained a firm hold. The combination of growing concerns about the impact of the era's new media and growing concerns about emotional stimulation—both of which contributed to each other—encouraged a range of thinkers to locate emotion in media technologies themselves as well as in a decidedly technologized version of the human body. This perspective supported phonograph companies' claims about the emotional power of recording technology, even as it served media researchers seeking to solidify their position as legitimate scientists looking "objectively" into the electrical processes of both technologies and bodies. In its framing of the relationship between people and communication technologies, media physicalism suggested a variety of contradictory and often highly problematic positions, drawing connections between technological development and emotional civilization (with all its race, class, and gender consequences) and suggesting that the quality of a people's communication could be determined primarily on technological grounds.

The following chapters offer a history of the rhetoric of media physicalism, showing some of the ways that notions of assumed technological power get attached to ideas about emotional stimulation during the early twentieth century and then exploring the implications of the resulting positions. In analyzing the growing dominance of this rhetoric, I focus predominantly on the discussions and debates of people in positions of power. Journalists, clergy, educators, radio announcers, politicians, scientists, media producers, and others of similar authority have unique platforms from which to shape these debates. As these discussions make their way into newspapers, radio programs, laws, classrooms, and the design and marketing of technologies themselves, they become part of the wider rhetoric informing our technological and emotional lives.

In focusing on these elite positions, I do not mean to suggest that everyday people are somehow brainwashed into believing in some dominant understanding of emotion or technology—physicalist or otherwise. People need not believe in a dominant emotional or technological rhetoric for it to take a toll on them. Instead, these elite discussions create a set of dominant cultural expectations against which our own emotional and technological displays are likely to be judged. This dominant culture can also set some very practical limitations on our communications. If a technology is designed in a specific way, based on the perceived benefits it might provide, it may be difficult if not impossible for us to use it in other ways. The fact that mass-produced radio receivers were not designed to transmit, barred most users from one kind of participation in broadcast culture. Although a group of technologically savvy amateurs found ways to subvert this limitation, most radio users remained more passive listeners.

Similarly, if media producers believe that a particular kind of music stimulates consumers' emotions in a way that is beneficial to their company, they are likely to produce that product rather than others. The Frankfurt School scholar Theodor Adorno has often been accused of an elitist dismissal of the everyday listener because of his critiques of the mass-produced nature of the popular music industry. While Adorno certainly does not waste love on the popular music audience, his critiques of the culture industry also point to wider problems of which the listener is a victim.[35] If the industry is driven to produce standardized music, which is believed to produce standardized emotions beneficial to the financial gain of recording companies, then listeners will have their musical choices seriously curtailed. Listeners do not need to be passive consumers or brainwashed drones to experience the limiting effects of the culture industry; they feel them every time they look for a media product that is not available to them.

In fact, it might be said that the elite decision makers are the ones most directly "brainwashed" by the culture's ideas about emotion and technology. They are generally the most engaged in debating these ideas and tend to have the most at stake in how they play out. Without a doubt, and as the following chapters illustrate, media experts can become quite fervent in their defenses of and attacks on media technologies. Despite their presumed status as outside observers, these experts are rarely free from the anxieties and ideologies at work in the broader culture. Established norms about technology and emotion will guide how media companies, media scientists, communication policy makers, and others think about their respective work and will have powerful secondary effects on the culture at large. Media physicalism, in all its paradoxes and contradictions, exercised just this sort of power in the early twentieth century, and it continues to do so in our own time.

The following chapters explore some of the complexities of U.S. understandings of communication, technology, and emotion that play into a larger rhetoric of media physicalism. Chapter 1, "Conflicting Feelings: Technology and Emotions from Colonial America to the New Age of Communication," offers some historical and theoretical background to set the groundwork for the more focused, early twentieth-century case studies that follow. Drawing together Leo Marx and Peter Stearns, I trace America's rhetoric of the technological sublime with a particular attention to its intersection with the history of emotion. A belief in the uniqueness of the new American frontier played a fundamental role in the country's eighteenth- and nineteenth-century technological and emotional rhetorics. I end the chapter by discussing how this larger history fed into the technological and emotional rhetorics of the early twentieth-century United States. Similarly to our own moment, a number of people described this period as high-speed and chaotic, with automobiles and other developing technologies contributing to this perspective. This sense of intensity created a belief in the uniqueness of the period that encouraged the strong rhetorics of sublimity that pervaded a number of discussions of technology and emotion. It was just this sense of emotional and technological power that set the stage for the era's growing wave of media physicalism.

Chapters 2, 3, and 4 explore the rhetorics of technology and emotion around specific early twentieth-century communication technologies. Each of the basic technologies I address in these chapters was both commercially available to everyday consumers and employed by social researchers and other scientists in their laboratory analyses. Likewise, both the popular and scientific discussions of these technologies tended to foreground their abilities to capture and transmit emotion, and the marketing of these technologies regularly employed the language of science. Finally, and as a result, each of these technologies became associated with education, being marketed for and used in school curricula, as well as a variety of self-help or correspondence course formats. The marketing, discussion, and dissemination of these technologies contributed to larger cultural attitudes about technology and emotion, driven by a general faith in the technological sublime as well as specific concerns about race, class, or gender that inflected each individual discussion.

Chapter 2, "Touching Images: Stereoscopy, Technocracy, and Popular Photographic Physicalism," explores the marketing, sale, and scientific use of stereoscopes from the 1890s to the 1920s. These three-dimensional viewers had been developed in the nineteenth century, but found a new popularity in the early twentieth-century United States. While stereoscopes may be largely unknown today, their early twentieth-century success offers a fascinating snapshot of the power of media physicalist rhetoric. Despite the stereoscope's nonelectronic, old-technological status, U.S. companies worked hard to establish its high-tech, sophisticated nature. The power of stereoscopic technology, the argument went, came directly from its capacity to transmit not only images but *feelings.* Used appropriately, these companies promised, the stereoscope would stimulate sentiments of nationalism, whiteness, and middle-class cultural capital appropriate to the new technological age. At the same time, stereoscopy began to be used in geography, medicine, astronomy, optometry, and a range of other scientific fields. While there was much debate about the scientific reality of the stereoscopic effect, many of these scientists reiterated the technological power of the stereoscope, giving a force to the physicalist claims made by their commercial developers. Drawing on these scientific studies, the two largest stereoscope companies, Keystone Viewing Company and Underwood and Underwood, created sets of stereoscopic slides, books, and other related material as kinds of high-tech self-improvement courses. Using a high-tech stereoscope, these companies promised, could transform someone into a more civilized person by cultivating his or her emotions in a range of powerful ways.

Chapter 3, "Electrifying Voices: Recording, Radio, and the New Friendly but Formal Speech," explores the impact of recording and radio broadcasting on speech delivery, teaching, and research. The period of the 1910s to the early 1930s saw the speech discipline replace the highly emotional practice of elocution with the more supposedly emotionally restrained practice of public speaking. This change reflected both concerns about emotional control encouraged by the new media age and the use of various technologies by speech teachers, scientists, and others. This new speech was to be both conversational and highly polished, illustrating concerns inspired by the new possibilities of amplification. Discussions of broadcast speech figured prominently in these debates, as the radio announcer became an exemplar of the new mass-mediated subjectivity. The radio announcer enacted the emotionally controlled life with technology to which each American was supposed to aspire.

Chapter 4, "Projecting Emotions: Motion Pictures, Social Science, and Emotional Self-Control," explores a set of scientific and popular concerns about motion pictures and related recording technologies. This chapter focuses primarily on the work of Christian Ruckmick, a psychologist in Carl Seashore's Iowa laboratory. As a colleague of Seashore, Ruckmick used apparatuses such as the psycho-galvanometer—a device that used film to record subjects' emotional responses—to analyze a range of communication phenomena. Ruckmick was the primary researcher on one of the two Payne Fund motion picture studies completed at Iowa, and the most explicit study of emotion among the thirteen final Payne Fund monographs. For this research, Ruckmick hooked up a group of children and adults to the psycho-galvanometer, had them watch motion pictures, and then recorded their emotional reactions to various scenes. This research, like other studies taking place at Iowa and in much of the wider scientific community at the time, illustrated an interesting array of anxieties both within the general culture and among early twentieth-century social scientists. The same technological power that made motion pictures dangerous stimulants of emotion made the recording apparatuses of the laboratory ideal for emotional analysis. These technologies also allowed Ruckmick and his colleagues in the social sciences to project an image of an emotionally controlled, scientific objectivity that sought to highlight their own immunity to the emotions of the new media age.

Departing from these early twentieth-century case studies, chapter 5, "Connecting Centuries: The Legacies of Media Physicalism," considers some of the ways these earlier attitudes about emotion and technology have carried over into the early twenty-first century. Although there would be a series of ebbs and flows in both social scientific and popular attitudes toward technology and emotion, the physicalist outlook of the early twentieth century became embedded in a range of institutions and practices. The Communications Act of 1934 remained in force until 1996, and its replacement, the Telecommunications Act of 1996, reiterated much of the earlier act's stance on technological and emotional matters. Likewise, and as I have already suggested, many of the utopian and dystopian concerns that greeted the radio and phonograph have surrounded the new media of the digital age as well. The social scientists and other thinkers of the twentieth century had established an American attitude toward new media in general. Media technologies were a set of technological and emotional transmissions directly impacting audience members, *who were themselves but a collection of technological and emotional impulses.*

Chapter 5 considers some specific early twenty-first-century examples of the rhetoric of media physicalism, showing how the same concerns about technological sophistication and emotional stimulation that characterized the early twentieth century are still at work today. Looking at such popular works as Steven Johnson's *Everything Bad Is Good for You: How Today's Popular Culture Is Actually Making Us Smarter,* and Nicholas Carr's book *The Shallows: What the Internet Is Doing to Our Brains,* as well as more academic work on media and emotion, I show how a persistent thread of media physicalism unites these positions to each other, as well as to the early twentieth-century work I discuss. In the second part of the chapter, I use research in the philosophy of mind to both critique media physicalism as an approach and to further explain the theoretical and philosophical premises that underlie the arguments and analyses throughout this book.

It is clichéd to say that there is some benefit in "talking about our feelings." While this book emphasizes the importance of this kind of talk—broadly conceived—it is ambivalent about its relative benefits. Certain kinds of talk about technology and emotion, such as those that make up the rhetoric of media physicalism, can have a range of constraining effects on a culture. Still, I hope to demonstrate that *talking about this talk*—thinking about how a culture expresses and negotiates its celebrations and concerns about mediated emotion—can help us see possibilities occluded by less thoughtful or reflexive kinds of conversations. In taking this approach, this book shares common themes with work in what is called the rhetoric of science or, more directly, the rhetoric of inquiry. The rhetoric of science, building on the research of such writers as Thomas Kuhn, explores how research practices, styles of writing, and larger intellectual paradigms structure the thinking of the natural sciences. The rhetoric of inquiry, as a group of scholars at Seashore's University of Iowa would eventually call it, expands this perspective to include the human sciences such as psychology and sociology as well as such areas as law, economics, and political science.[36]

Scholars in both the rhetoric of science and the rhetoric of inquiry have offered useful ways of analyzing a range of established academic disciplines, attempting to understand, for example, how the paradigms of a field like biology structure what scientists know about the body as well as how biological information is communicated from experts to the public. However, the early twentieth-century "media researchers" I discuss in the following chapters did not participate in a common, formal discipline. They were a loose coalition of psychologists, sociologists, speech teachers, and others trying to make sense of the emerging communication technologies of their period. A field of media or communication research would not be formalized for several decades. Likewise, and as a result, the lines between academic and public understandings of the media were not as distinct as they might otherwise have been. There was no well-established scientific language about the media to be translated for the public, although many researchers were trying to establish one, and a number of media producers were trying to take advantage of it as they did. The rhetoric of media physicalism was an emerging blend of cross-disciplinary academic perspectives, the promotional claims of commercial media producers, and larger public discussions of technology and emotion.

For like reasons, the rhetoric of media physicalism has an uneasy fit with one of the central focuses of rhetoric of science and rhetoric of inquiry scholarship: argument. One of the early goals of both approaches was to demonstrate how even ostensibly objective sciences were, in fact, making arguments and creating persuasive appeals. Chemical formulas were not simply abstract renderings of a hidden reality; they were a certain kind of persuasive attempt to win people to a chemical reading of the world. Charles Bazerman, an important figure in the rhetoric of inquiry, has used this persuasion-focused approach as part of his own "rhetoric of technology." His book *The Languages of Edison's Light* seeks to understand how Thomas Edison, as scientist and businessperson, gave his light meaning and sold it to the larger culture. Bazerman does not cast Edison as some all-knowing, all-powerful persuader of the public, however; rather, he suggests that the inventor had to negotiate with the rhetorical power of the electric light itself. As Bazerman explains, "The night lit up at the flick of a switch argued for itself, electrocution of beast and man signified electricity's terrifying power, and regular delivery of light was one means of persuading consumers to pay their monthly electric bills." Still, the fact that "Edison was savvy enough as a rhetorician to use all these material arguments" explains for Bazerman much of his success.[37]

In the examples of media physicalism that I address, there is no clear rhetorician or group of rhetoricians offering a unified, persuasive message. Certainly the people I discuss engaged in various sorts of persuasion. Carl Seashore attempted to advocate for a certain psychology of music and, eventually, sell his own test of musical talent; stereoscope companies attempted to sell a largely outdated technology as a high-tech one; speech teachers argued for a new, supposedly scientific understanding of public speaking; radio producers sought to sell audiences a specific version of a presumably personable announcer. The physicalist understanding of communication technology and emotion that emerged from these messages and interactions was largely a by-product of these other, more explicit goals. None of the thinkers I discuss directly advocated a physicalist stance, at least in name; however, as I will demonstrate in the following chapters, both individually and collectively, their messages and practices gave this perspective a particular cultural power and even a sense of inevitability.

As a way of addressing these complexities, I approach media physicalism through a perspective that Jenny Edbauer and Nathaniel Rivers and Ryan Weber call *rhetorical ecology*.[38] According to Edbauer, rhetorical scholarship has tended to focus on concrete "rhetorical situations," imagining a clearly delimited time and space in which rhetorical meanings are created. While this might make sense when analyzing a speech that takes place in a specific location and moment in time—and even here rhetorical theorists have raised questions[39]—this view cannot account for some of the more complicated interactions to which contemporary rhetorical criticism has turned. Edbauer argues that such work demands a different approach that recognizes that "the intensity, force, and circulatory range of a rhetoric are always expanding through the mutations and new exposures attached to that given rhetoric." Recognizing these energies and mutations, rhetorical ecology "reads rhetoric both as a process of distributed emergence and as an ongoing circulation process."[40]

This book undertakes a kind of *rhetorical ecology of inquiry*, exploring the distributed emergence and ongoing circulation of media physicalism. The presumed sublime power of early twentieth-century communication technologies and a parallel concern for emotional control were a "shared contagion"[41] inflecting the era's understandings of media and feelings more generally. Phonograph companies' claims that recorded music gave audiences a uniquely powerful form of emotional experience were closely connected to educators' and others' fears about the emotional dangers of movies as well as to psychologists' arguments about the power of their own emotion-recording technologies. As a result, even when these various thinkers seemed to be strongly opposed to each other, they often implicitly conspired to build quite similar understandings of mediated emotion. Each of the individual voices I explore was thus both a product and producer of the larger rhetorical ecology of media physicalism. Even as they responded to extant ideas about the emotional power of media technology, they pushed them forward in the form of product advertisements, popular articles, scientific studies, and media policies.

In exploring these discussions as a rhetorical ecology, I work to stress various sorts of connections across the topics and people I address. The fact that Carl Seashore's laboratory used and studied stereoscopes and that Seashore was mentioned in various stereoscope company publications evidences the often quite concrete relationships between media researchers and media producers during this period. Indeed, Seashore and his colleagues appear across the various early twentieth-century chapters, as do other important figures in the history of media research, such as the psychologist E. B. Titchener, who was the mentor of Christian Ruckmick before he joined Seashore's department. A number of academic institutions also appear across these chapters, including such universities as Iowa, Wisconsin, Ohio State, and University of Chicago, all of which were active in the study of communication during the early twentieth century.

In making these connections within and across the academy itself, I explore some important and well-known ideas in the history of the social sciences—such as psychology's rejection of philosophy and its abandonment of introspection—with an eye to their importance for the history of mediated emotion. As I discuss in more detail in chapter 4, introspection was a research method in which people were asked to reflect upon their own internal bodily and psychical processes. The fact that psychologists studying early twentieth-century media were also instrumental in rejecting introspection is an important feature of the rhetorical ecology of media physicalism. The view that emotions were primarily bio-technological processes made them seem both dangerous, for media users and media researchers alike, and below the surface of people's introspective perception. Not surprisingly, and as I explain in chapter 4, by the 1930s, various forms of emotion-recording technologies came to replace introspection in most psychological labs. While this book is not a history of psychology per se, the connections I draw suggest that some central concepts in psychology's history may be intricately connected to broader cultural ideas about media and emotion. By the same token, although I won't spend much time on physicalism as a formal philosophical position, my arguments about the less philosophically sophisticated position of media physicalism may well suggest that anxieties about emotion and technology undergird physicalism more generally—that it, too, emerged as part of this particular rhetorical ecology.

By tracing these connections within the academy and showing how the academy's ideas connect to wider discussions of the media, this book also offers a historical perspective on the emergence of *administrative media research*. As I mention above, Paul Lazarsfeld identified this model of research in the 1940s, but Carl Seashore and a range of other social scientists engaged both the government and corporations in administrative fashion almost from the beginning of their research into communication. This had a powerful influence over the kinds of studies these researchers undertook, which typically dealt with questions about the shaping of attitudes or the emotional impacts of technologies that were especially interesting to business executives or administrative officials. These administrative impulses were an important part of the early rhetorical ecology of media physicalism, whose technological and physiological perspective was especially well suited to corporate and governmental concerns. In an important sense, this book demonstrates, administrative communications research and media physicalism have intimate connections to each other, at least in the American context. To the extent that media physicalism still dominates American thinking about the media, it does so alongside a still troubled set of administrative tendencies.

Of course, it is easy to assume that arguments about introspection or scientific studies of media use are disconnected from popular ideas about the media. In exploring how scientists, producers, educators, clergy, and others engaged with the new media are part of the same rhetorical ecology, I hope to show the wide reach and broad consequences of our thinking about emotion and media in general, as well as of media physicalism in particular. Although I place the emergence of the media physicalist position in the early twentieth-century United States, it built upon a still wider set of rhetorical ecologies (including the earlier American ideas about emotion and technology I address in chapter 1). Just like concerns about emotion and technology more generally, certain aspects of media physicalism may have been at work in Socrates's concerns about writing or elocutionists' celebrations of print. Still, the early twentieth-century confluence of the emergence of broadcasting, the adoption of a highly commercialized, market model for thinking about the media, and American social scientists' particular perspective on these emerging phenomena inflected the rhetorical ecology of media physicalism in some especially interesting and problematic ways. Indeed, I will argue that media physicalist thinking has become a central component of how a large number of American scientists, policy makers, journalists, and others make sense of communication, which, beginning in the early twentieth century, came to be seen predominantly as a technological process whose consequences were primarily about informational and physiological *effects*.

In taking aim at these ideas, this book takes seriously James Carey's claim that twentieth-century American thinking about the media has been dominated by a "transmission model" that stresses technology, efficiency, and impact at the expense of the interaction, dialogue, and community encompassed in the "ritual view" that Carey favored. The emphasis on transmission has resulted in communication teaching and research that focuses on how someone might best affect another person—or avoid being affected—rather than on how people join together in communal meaning making.[42] In offering the history that it does, this book suggests one powerful source for the dominance of the transmission model of communication—the narrow views of technology and emotion encompassed in media physicalism.

Exploring media physicalism's claims about technology and emotion demonstrates the powerful consequences of a technologically centered, transmission view of communication. If we assume that how people *feel* about a particular piece of media is primarily a question of how their body responds to it physiologically, or if the quality of a message depends primarily on its technological features, we can avoid asking some tough questions about the ethics of communication. In fact, as I will demonstrate in the following chapters, in their adoption of media physicalism, early media researchers explicitly tried to avoid a series of social, historical, cultural, and ethical questions.

However, questions about who gets access to the media and how, about what kinds of values we want our media to uphold, about the intricate role of media in the creation of culture, and about all the various ways we might understand quality communication don't simply disappear when media researchers take up emotion-measuring technologies. To take one powerful and important example, because claims about technological power and emotional control inevitably have class, race, and gender assumptions built into them, media physicalism offers a quite problematic politics of identity, even as it suggests that such questions are not relevant to media use or research. Undoubtedly, media physicalists of both the early twentieth and early twenty-first centuries repeatedly make claims about society and culture—for instance, arguing that people are becoming too emotionally keyed up, or that, as a culture, our lives are faster, better, or worse, than the lives of people in the past. However, in tying these claims to the bodies and brains of individual people, media physicalism ultimately fails to explore what should be central components of the social aspects of mediated emotions—how a group of people envision themselves, their technologies, and their emotions, how this vision is communicated, and the consequences of this vision for those who do and do not fit within it. In seeking to understand and challenge the rhetoric of media physicalism, this book hopes to bring these important questions to the fore.

1

Conflicting Feelings

Technology and Emotions from Colonial America
to the New Age of Communication

Benjamin Franklin was one of the first American media theorists. A printer, newspaper publisher, and postmaster, Franklin produced and thought about a range of media forms. In his frequent discussions of "communication," however, he primarily had something else in mind. In explaining an experiment with electricity, Franklin instructed his readers to "place a thick piece of glass under the rubbing cushion to cut off the communication of electrical fire from the floor to the cushion."[1] Similarly, in an explanation of a rudimentary battery made from a bottle, Franklin wrote that "the Equilibrium cannot be restored in the Bottle by *inward* Communication, or Contact of the Parts."[2] In a discussion of how the lightning rod he designed could help a church, Franklin offered that "a sufficient metallic communication between the roof of the church and the ground" needed to be established.[3] For Franklin, *communication* was primarily an electrical interaction between physical objects rather than an exchange between people.

Franklin's use of the term "communication" reflected a common sense of the word in his time, though one that was even then beginning to change. As John Peters has explained, "The concept of communication as we know it originates from an application of physical processes such as magnetism, convection, and gravitation to occurrences between minds."[4] The seventeenth-century fascination with electricity of which Franklin was an important part created a like interest in various other kinds of connectivity. If metal objects could develop a magnetic attraction to each other, then what about human beings? The fact that we can still speak of a person's "magnetic" personality testifies to this earlier understanding of communication as electrical attraction. Our sense of communication is rooted in an idea of powerful electrical impulses drawing one body to another.

The concern with technology and emotion that developed in the early twentieth century owed much to this earlier idea of communication. The modern concept of communication as sharing of information or meaning matured alongside a range of technologies that themselves connected bodies via electricity. In harnessing electricity as it did, Morse's telegraph—patented in 1837, not quite fifty years after Franklin's death—embodied and magnified the magnetism of the new age of communication. If in Socrates's time the alphabet had caused concerns about emotional stimulation and both connection and disconnection, the electrical alphabet of Morse's code would seem to transmit emotion itself. The telegraph carried the magnetism of one body to another, via its own magnetic wires.

In exploring this early history, I will argue that, at least in the American context, the rhetoric of the technological sublime and the history of emotion need to be read as parallel parts of a larger rhetorical ecology. The presumed power of electrical technologies, especially the telegraph, evoked great interest in and concern about emotion. In the nineteenth century, a number of commentators celebrated the telegraph as a great emotional unifier—a national heart. The more its wires multiplied, however, the more people worried that it was causing a form of emotional overstimulation. These ideas moved hand in hand. The more powerful and omnipresent communication technologies such as the telegraph seemed to be, the more the emotion they seemed to transmit was imagined as a set of electrical impulses in need of control. The more emotions were seen as electrical impulses in need of control, the more powerful these communication technologies seemed to be. This circular logic eventually culminated in the early twentieth-century position that I call media physicalism.

This chapter traces the history of this developing logic from the founding of the country to the early twentieth-century period that grounds the next several chapters. In comparison to later periods, early Americans saw emotional life as a social and public good. Feelings were to be expressed publicly, and they formed an important component of people's bonds to one another. Toward the turn of the twentieth century, the continual growth of communication technologies, combined with an increasingly urban life, created a range of anxieties about the speed and amount of stimulation—emotional, informational, and otherwise—that a person could handle and called into question the earlier, more public conception of emotion. Not only telegraphs and radios, but such new technologies as the automobile created what many saw as a new era of hyperemotional stimulation and "speed mania." This linkage between technological development and emotional stimulation created a central anxiety for thinkers of the period: *What did civilization mean if the very technologies that advanced it also created emotions that were dangerous to its development?* This apparent paradox weighed heavy on a whole range of arguments about the place of technology and emotion in human life, suggesting that technologies were both the causes of—and solutions to—emotional overstimulation; enforcing the need for administrative experts to guide the technological needs of the public; and depicting the average citizen as an overwhelmed, hyped-up addict of the era's technological emotions.

God's Lightning and the National Heart

For nineteenth-century Americans, perhaps nothing embodied the powerful combination of technology and emotion quite as strongly as the telegraph. The first stanza of a poem by Elizabeth Barnard entitled "The Atlantic Telegraph," published in 1883, imagined the telegraph's wires as a great emotional unifier:

> Peerless theme of glad emotion
> Linking national hearts in one;
> Through this nerve across the ocean
> Thrills the triumph newly won![5]

Massachusetts senator George F. Hoar likewise celebrated the shared sentiment made possible by the telegraph. In one public address, Hoar observed that "every speaker and every auditor knows how an emotion is multiplied by the size of the audience that feels it." Someone might tell a joke to a neighbor "which will hardly create a smile." However, "say the same thing to a great audience of three or four thousand people, and in every man's heart that feeling is multiplied and intensified by the knowledge that the same feeling is experienced by every other person."

Addressing the assassination of President Garfield, Hoar claimed that "science, the telegraph and the press enabled the emotion of human sorrow, at the time of Garfield's funeral, to be felt over the entire civilized world." Because of the sharing of emotion made possible by the telegraph, with Garfield's assassination, "a poor, feeble fiend shot off his feeble bolt; a single human life was stricken down; and, lo, a throb of Divine love thrills a planet!"[6]

This same telegraphic sharing of feelings created worry as well. According to a letter in the *Philadelphia Medical Times* from 1883, the emotional stimulation of the telegraph could do harm to the human body:

> All day long there is the telegraph boy with his sharp summons and the emotion which is inseparable from the nature of the message sent. When a man only got his letters in the morning he was pretty safe from surprises for the rest of the day; but with the telegraph he has no remission from anxiety and is on the tenter-hooks all day long. . . . What chance have the assimilative organs, so intimately related with the emotions, of preserving their even way amidst such tumult and disturbance?[7]

In a like vein, an article entitled "Intellectual Effects of Electricity" argued that the telegraph's "constant excitements of feeling unjustified by fact . . . must in the end, one would think, deteriorate the intelligence of all to whom the telegraph appeals."[8] Still another writer suggested that the telegraph had potentially harmful effects on people's emotions, in that it "searches every nook and corner of the world every day, dragging into light, not only every crime that is committed, but every disagreeable feature of human society."[9]

Summing up some of these contradictory positions, a poem that celebrated the telegraph as both "grandly and simply sublime" and a "sensitive link" binding people in mutual feeling also warned of the potential dangers of these connections:

> But ye must watch it in good sooth
> lest false fever it swerve
> touch it in tenderest truth
> as the world's exquisite nerve.[10]

Exquisite nerve and national heart, the telegraph embodied for these commentators a particularly powerful—and contradictory—technological emotion.

These contrasting positions on the value of the telegraph's emotional unification make sense in the context of America's larger histories of technology and emotion. As Leo Marx has explained, early American confrontations with technology were dominated by what he terms, following the historian Perry Miller, "the rhetoric of the technological sublime."[11] Because of its vast, uncultivated land areas, Marx argues, from the beginning of the age of discovery America seemed the perfect setting for the classic "Virgilian mode." In this archetypal formula, a good shepherd would "withdraw from the great world and begin a new life in a fresh, green landscape."[12] The mythic American frontier proved for many a powerful counterpoint to the supposed civilizing influences of Europe and suggested a mode of sublimity unique to the new continent. Hawthorne's Sleepy Hollow, Thoreau's Walden Pond, Jefferson's pure and innocent Virginia, and a myriad of other literary and political images on which Marx draws seemed to celebrate a pure American landscape untouched by modern life. Influenced in various ways by the European aesthetics they were presumably escaping, however, even as these new settlers praised the American wilderness, they fantasized its transformation into the more cultivated, civilized "middle-landscape" of the garden. Jefferson's ideal citizen was the yeoman farmer or "husbandman" who turned the chaotic wilderness into a more ordered, productive space for the cultivation of crops. The ideal landscape assumed a middle ground between some primitive, untouched nature and a more cultured, civilized one.

The example of the locomotive provides Marx with a strong example of this general rhetoric. Although such writers as Hawthorne and Emerson initially decried the railroad for the ways that it disturbed the bucolic American prairie, with increasing frequency, people eventually began to celebrate its technological mediation of the landscape. "There is a special affinity between the machine and the new republic," Marx observes, because "the raw landscape is an ideal setting for technological progress."[13] Wolfgang Schivelbusch notes a similar ambivalence in the European reception of the railroad. While those people accustomed to the slower travel of the horse-drawn carriage were often critical of the train's rapid movement through the landscape, others began to champion the uniquely technological view of the prairie made possible by locomotion. These spectators did not see "a picturesque landscape destroyed by the railroad." For them, the train had created a new landscape, to be taken in through a series of high-velocity "glances."[14] Americans moved still more quickly to this more celebratory view of railroad travel. However disturbing the railroad might have seemed to the presumably pristine American landscape, in its raw power to consume the countryside it also appeared to many as a natural symbol of a developing ideology of technological progress.

In discussing the technological sublime, Marx, like Miller before him, tends to employ the term "sublime" in a fairly mundane sense—as, say, wonder or excitement—using it primarily to describe various *celebrations* of technology. However, the ambivalence that greeted both the telegraph and the railroad suggests that a notion of sublimity drawn more clearly from the work of Edmund Burke or Immanuel Kant might be more appropriate. In his classic eighteenth-century discussion, Burke argued that a person experienced the sublime when faced with something that evoked a sense of vastness, magnificence, power, infinity, terror, or another sensation of astonishment. For him, "delightful horror" was "the most genuine effect, and truest test of the sublime."[15] Kant likewise explained that when someone experiences a sublime object, "the mind is not simply attracted by the object, but is also alternatively repelled thereby." As a result, "the delight in the sublime does not so much involve positive pleasure as admiration or respect, i.e., merits the name of a negative pleasure."[16] The sublime, as understood by Burke, Kant, and many of their eighteenth-century counterparts, was a feeling of fearful wonder that resulted from the confrontation with some terrifyingly awesome object.

That the telegraph could be both celebrated as a "peerless theme of glad emotion" and condemned for its "constant excitements of feeling unjustified by fact" suggests that it was greeted with just this sense of delightful horror. This grew in part from its status as *electrical* communication, electricity itself being met with its own powerful rhetoric of sublimity. Edmund Burke had early on suggested that attending to "the last extreme of littleness" could evoke a sublime experience. When humans considered objects of a "diminishing scale of existence," they would "become amazed and confounded at the wonders of minuteness."[17] Electricity had just this sense of wonder. In 1874, one minister claimed that "since electricity has become known (in part), it furnishes a far more forcible symbol of spirit, and even of divine power." In fact, he contended, "it has many of the attributes of those spirits which the Almighty makes his messengers, the flame of fire which he makes his ministers."[18] Still another writer argued that "from electricity, which is the invisible body of God, have emanated all the visible substances that constitute globes, and from the fullness of his spirit have emanated all life, form, and motion."[19] The sublime power of electricity was one with the sublime power of God.[20]

By harnessing electricity for the purpose of communication, the telegraph seemed to give the human voice a godlike reach. In an early history, Charles Briggs and August Maverick called the telegraph "a perpetual miracle, which no familiarity can render commonplace." Given the telegraph's miraculous character, they asked, "For what is the end to be accomplished, but the most spiritual ever possible? Not the modification or transportation of matter, but the transmission of thought."[21] An article describing how the telegraph was used to spread information about crimes claimed that "God's lightning pursuing murder has become a true and active thing."[22] According to James Carey, whose own concept of the "rhetoric of the electrical sublime" built upon Marx's ideas, the telegraph entered "American discussions not as a mundane fact but as divinely inspired for the purpose of spreading the Christian message farther and faster, eclipsing time and transcending space, saving the heathen, bringing closer and making more probable the day of salvation."[23] Owing to this mysterious, transcendent understanding, the telegraph also became attached to a range of psychic practices such as spiritualism and mesmerism.[24] The mysterious power of the telegraph promised, in the words of Gardner Spring, "a spiritual harvest because thought now travels by steam and magnetic wires."[25]

That not just thought, but *emotion* could be carried by the telegraph was a central component of its apparently sublime power. The train was delightfully horrifying for how it travelled the vast American landscape, "annihilating space and time."[26] The country's expansive railroad tracks testified to the force of technological progress and the inevitable taming of the raw prairies of the American frontier. Telegraph wires covered much of the landscape as well, and, in fact, often ran parallel to railroad tracks as telegraphic signals were used to regulate time between stations and provided other communications important to train travel.[27] Whereas the railroad transported people, however, the telegraph, as exquisite nerve and national heart, seemed itself a technological embodiment of a certain kind of personhood, stretching Americans' nervous systems to the far reaches of the landscape. In another annihilation of space and time, the telegraph could seize sentiments from any corner of the country and quickly transmit them to all the others. To its critics and celebrants alike, the telegraph was a sublime technology with the ability to transmit sublimity itself.

The concept of a national heart had a particularly important meaning during this time period, in that the highly social idea of the passions was still an important part of the public consciousness. Before the more narrow idea of emotions that is explored in the following chapters came to prominence, emotions could still be both public and powerful without necessarily being seen as dangerous. In fact, in certain circumstances a cultured response entailed extremely public, highly animated emotional displays. As one eighteenth-century writer explained, reason alone is not enough to guide a public because it "is like an old Man, full of Prudence and Sagacity, who judges excellently, but wants Vigour and Agility to act." The public required passion as well, for "Reason shews the Goal, and the Passions animate the Race."[28] An essay published in the *New York Magazine* in the late eighteenth century claimed that "passions are in the moral what motion is in the natural world," because "the passions animate the moral world."[29] According to these writers, the cultivation of passion played an important role in maintaining a healthy public.

A powerful, public conception of the passions played an important part in early American religious life as well. The religious revival commonly referred to as the Great Awakening, which began in New England in the 1730s and 1740s, placed a high premium on the experience and expression of emotion. One of the great champions of the revival, Jonathan Edwards, stood as a strong proponent of religious passion. His sermon "Sinners in the Hands of an Angry God," first delivered in Enfield, Connecticut, on July 8, 1741, testified to the powerful terror through which human beings should view their Lord:

The God that holds you over the pit of hell, much as one holds a spider, or some loathsome insect, over the fire, abhors you and is dreadfully provoked; his wrath towards you burns like fire; he looks upon you as worthy of nothing else but to be cast into the fire; he is of purer eyes than to bear to have you in his sight; you are ten thousand times so abominable in his eyes, as the most hateful and venomous serpent is in ours.[30]

The fervent preaching of Edwards and other ministers involved in the Awakening spawned a popular religious movement. On at least one occasion, after hearing Edwards preach, members of his audience gathered with him in a private house in which could be heard not only "sobs," but "Groans & Screaches as of Women in the Pains of Childbirth" and "Houlings and Yellings, which to Even a Carnal Man might point out Hell, & Convince him that Concience [sic] let loose."[31] Edwards championed both his own and his congregation's impassioned emotions. Provided he spoke the truth, he held, "I should think myself in the way of my duty to raise the affections of my hearers as high as I possibly can,"[32] even if that meant "speaking terror."[33] Edwards and his followers viewed the public sharing of highly impassioned speech as a fundamental part of religious experience.

Religious sentiments were not the only ones to be shared in public. Despite the presumably private and intimate communications they entailed, public love letters were common among eighteenth-century Americans, as Nicole Eustace has demonstrated. Caught between earlier practices in which young people's parents were expected to select their marriage partners based on various economic considerations, and later ideals that stressed romantic love, eighteenth-century suitors often addressed their impassioned letters not to their intended, but to one of his or her family members. When the Philadelphian Henry Drinker expressed his powerful love for Betsy Sandwith, he addressed his letters to her sister, Mary. Suitors performed these public displays of affection in order to demonstrate their commitment to a range of familial and social norms. The more willing people were to publicly express their love, the more devoted they seemed to be toward the various relations of power that supported the period's bonds of kinship.[34] Such powerful and public expressions of emotion could both demonstrate and secure one's social standing.

Other public expressions of emotion carried similar cultural power. For instance, in the late eighteenth and early nineteenth centuries, the public expression of tears could provide evidence of someone's virtue, good standing, and even masculinity. A poem entitled "A Tear," printed in 1799 in the *Rural Magazine*, champions both the sacred beauty and naturalness of crying. Stressing its spiritual power, the poem describes a tear as "a sweet drop of pure and pearly light" and a "benign restorer of the soul." The poem's final lines use the relationship between tears and the law of gravity to demonstrate their natural wonder and force:

> The very law which molds a tear,
> And bids it trickle from its source,
> That law presents the earth a sphere,
> And guides the planets in their course.[35]

A poem printed in an 1800 edition of *Weekly Museum* similarly claimed that "no radiant pearl which crested fortune wears" could "shine with such lustre as the tears that break / For other's woe, down Virtue's manly cheek."[36] Although there was some disagreement regarding the relative "manliness" of tears, a number of writers in this period, like this poet, found something eminently masculine about crying. The public shedding of tears, even by men, offered the kind of shared sociality and emotional unification that was the hallmark of the passions.[37]

The example of elocutionary speaking—the dominant speaking style taught at schools and universities during the nineteenth century[38]—demonstrates another context in which people were to celebrate and even cultivate the impassioned, public performance of emotion. Elocutionists recited poems and offered orations at a range of public functions, using elaborately designed gestures and vocal inflections. As William T. Ross explained, "Elocution is a means for artistic and intellectual culture. It is an accomplishment. It improves the conversational powers. To the possessor of the art, it is a solid satisfaction, and it enhances the enjoyment of society."[39] In developing a student's ability to display emotions appropriately, elocutionary study promised "the enlargement and elevation of human personality through the proper cultivation of the power of expression."[40]

In order to develop this elevated personality, elocutionists adopted a set of prescribed vocal and bodily practices assumed to evoke particular emotions. In the 1881 book *A Manual of Gesture*, Albert Bacon diagrammed and explained three kinds of hand movements—descending, horizontal, and ascending—each of which was divided into front, oblique, lateral, and oblique backwards positions.[41] Bacon explained the emotional force of these movements:

In the mechanical execution of gesture we employ straight lines and curves; as in geometry, to which the laws of gesture are referable. Straight lines, which indicate directness of thought, are employed to express bold, energetic and abrupt ideas. The curved lines are used in more calm and quiet states of mind, to express gentle and genial thoughts and emotions, and are also adapted to the boldest flights of oratory.[42]

An earlier writer explained that in order to achieve the correct emotional force when speaking, "we must be careful to let the stroke of the hand, which marks force, or emphasis, keep exact time with the force of pronunciation."[43] Through a mode of bodily and vocal control based on these well-defined practices, speakers were expected to engage in a highly theatrical, impassioned, emotional display.

This type of elocutionary practice performed what Peter Stearns has called a "Victorian style" of emotion. Although Victorian-era Americans have often been imagined as especially repressed, Stearns suggests that such interpretations ignore how this same culture often embraced

Figure 1.1

Right hand ascending oblique vertical, sacred, or sublime deprecation. Reprinted from Albert Bacon, *A Manual of Gesture* (Chicago: Griggs, 1881), 117. According to Bacon, this movement should accompany a phrase such as "Forbid it, Heaven!"

and celebrated the passions. Victorian-era interests "in strong wills, in romantic idylls, in evocative cemeteries, in tales of heroism, and, of course, in true, loving womanhood"[44] both evidenced and contributed to this often celebratory understanding of the passions. During this period, Americans regularly advocated both self-control and impassioned public expressions of emotions, as in the example of elocution. While this might seem contradictory, it resonated with larger concerns about decorum and taste that also inflected the period. Dominant cultural attitudes during this era often associated highly theatrical performances of the self with indications of a strong cultural upbringing. Such performances signified people's abilities to both take control over their bodies and to achieve some ideal of transcendent emotionality.

The phrase "dominant cultural attitudes" is fitting here, as not every class or group was equally included in this culture of emotion. Even among the educated classes, the emotions of women and men were to be cultivated differently. "So much depends upon the temper of women," claimed a writer in the *Boston Weekly Magazine*, "that it ought to be most carefully cultivated in early life; girls should be more inured to restraint than boys, because they are likely to meet with more restraint in society."[45] The uneducated were even more excluded from these emotional ideals. "The difference between a savage New Hollander, and a highly polished European, is as great as between animals of a distinct species," argued one early nineteenth-century writer. "By education the most powerful natural passions are either suppressed or strengthened."[46] Without proper education, people risked either succumbing to uncontrolled emotional outbursts or "drudg[ing] on through life with scarce any feelings or apprehensions beyond the present moment."[47] Of course, not every group excluded from this culture wanted to be part of it. From the perspective of one advocate of the working classes, these supposedly cultured classes were "burning under the unholy passion" of avarice, which "stifled every monition of conscience."[48]

Despite such disagreements and inconsistencies, the various edicts of this early American culture of emotional display had consequences throughout the newly formed nation, even for those outside the dominant culture. The religious revivalists of the awakenings, for instance, took a special interest in Native Americans, seeking to educate the young on the proper modes of emotional experience and performance. "The morning I last left the school," wrote one such missionary, "the two most forward boys, whom I use as interpreters, were bathed in tears under a sense of their sin." He imagined that "should they get religion they should be qualified to have easy access to the consciences of their friends."[49] Another writer commented that certain Native American traits were "so humane and generous as to produce in the civilized mind mingled emotions of astonishment and delight," observing that "it thus appears, that rude and uncultivated minds are susceptible of the finest sensibility, of the warmest attachments, of the most inviolable friendship."[50]

Recounting a story that echoed Jonathan Edwards's celebrations of religious terror, an article in the *United States Christian Magazine* reveled in an African American man's account of the awakenings of his own Christian sentiments after having read the Bible. When the narrator of the article asked the man about his conversion, he reportedly answered,

> Why Massah, I found that I had a very bad heart, Massah, a very bad heart indeed: I felt pain that God would destroy me because I was wicked, and done nothing as I should do. God was holy and I was very vile and naughty; so I could have nothing from him but brimstone and hell.

Moved by the man's confession, the narrator explained that the man had "entered into a full account of his convictions of sin which were indeed as deep and piercing as any I had ever heard of."[51] In the view of these advocates of the dominant emotionology, those groups outside the white, educated classes could still learn this culture of emotional experience, and thus be subject to its pressures. Those who failed or refused to do so faced consequences of their own. Claims about emotional propriety served to support a range of cultural practices and to condemn others—the weight of the latter falling heavily on those outside the dominant classes.

As these examples illustrate, although a range of countervailing forces inflected the emotional culture of late eighteenth- and early nineteenth-century America, certain dominant attitudes took form and persisted across these longer periods of time. In such mainstream practices as the awakenings, public love letters, elocution, and so forth, feelings were imagined as overwhelmingly public, social goods that needed to be cultivated and shared in order for their true power to be realized. This understanding is reflected in the eighteenth- and nineteenth-century preference for the word "passion" over "emotion." The passions carried a religious connotation that implicated them in a larger network of social relations. The word "emotion," particularly as it was understood by early twentieth-century Americans, emphasized more private, biological connotations distinct from this earlier American emotionology. The cultivated American of the era of the passions was to develop a range of extremely strong feelings and then channel them into impassioned public displays.

In tracing America's transformation from a social understanding of the passions to a more individual-focused idea of the emotions, the telegraph again provides a powerful example. Beginning in the 1870s, an interesting debate about the nature of telegraphic emotions arose around a series of legal cases in which customers sued telegraph companies for "mental anguish" caused by late or undelivered messages. In one precedent-setting case, on January 16, 1874, a woman died in Giddings, Texas, and one of her relatives sent a telegram to Austin, attempting to inform her son, C. O. So Relle, of her death and upcoming funeral. However, the telegram was not delivered quickly enough, and the young man sued Western Union over the mental anguish he claimed to have experienced as a result of missing his mother's funeral. Even after a lower court dismissed the case, the Texas Supreme Court held that "it appears to us that the natural consequence of a failure to promptly transmit and deliver a message like that in this case . . . is to produce the keenest grief incident to a disappointment." They found that So Relle had a valid claim and could recover the $50,000 he requested for his suffering.[52]

The *So Relle* case set a strong national precedent during the 1880s and 1890s. In an Indiana case, a man whose wife was about to die telegraphed his sister and brother-in-law to ask them to join him at her deathbed. As a result of a delay in the message's delivery, however, they were not able to arrive before she passed away. According to the Indiana Supreme Court, "by reason of their absence, and of the great desire the [man's] wife had expressed to see them before her death," the man had "suffered great uneasiness, anguish, and anxiety of mind."[53] In a North Carolina case, a young woman was left stranded at a train station because the telegraph company failed to deliver a message that her father had sent, asking a friend to pick her up. As a result of the ordeal, including the daughter's apparently traumatic taxi ride arranged by the "colored matron" of the train station, both the daughter and father claimed that they had suffered mental anguish.

Despite a lower court's ruling, the North Carolina Supreme Court found that both the father and daughter had legitimate claims.[54]

In keeping with the public understanding of emotion that was still alive at this time, these courts argued that telegraph companies had a public responsibility to the emotions of their customers. From the courts' standpoint, telegraph operators themselves played an especially important public role in the electronic dissemination of emotion. Because they had to *read* people's messages in order to transmit them, the Texas and other courts believed that telegraph operators had an empathetic responsibility to the emotions of their customers. When these operators received messages such as "Billie is very low; come at once"[55] or "my wife is very ill, not expected to live,"[56] they were expected to understand the emotional importance of the message and then to treat it with an appropriate level of concern. As the Indiana court mentioned above explained, when receiving such messages, a telegraph company was "bound to know that mental anguish might, and most probably would, come to some person in case it failed to act promptly in transmitting and delivering the dispatch."[57] Telegraph companies' responsibilities did not amount to simply transmitting a message. They also played an important role as public arbiters in people's emotional connections.

Sustaining the general rhetoric of the period, these cases also highlighted the sublime power of telegraphy. In deciding *So Relle*, the Texas court had relied upon Thomas Shearman and Amasa Redfield's *Treatise of the Law of Negligence*, which had likewise argued that telegraph companies should be held responsible for the mental anguish caused by undelivered messages. Shearman and Redfield based much of their position on the unique features of telegraphic communication. As compared to other kinds of business transactions, when people spent money on a telegraph message they were entering into a kind of transaction that was "*sui generis.*" Shearman and Redfield did their best to capture what they saw as the sublime mystery of the telegraph:

> The message must be put upon the wire by a series of longer or shorter pressures of the operator's finger. If he keeps his finger down a second too long or raises it a second too soon, he will destroy the meaning of the message, as will be seen when we describe the form in which it arrives at the other end of the line. The electricity which is let loose by the operator's finger then

passes along the wire, and delivers a series of shocks at every office, corresponding in length with the fingering of the operator. The particular operator who, by a peculiar signal, has been warned to prepare for messages, then listens to a series of snapping sounds, unintelligible to untrained ears, but which he is able to translate into letters of the English alphabet.[58]

For Shearman and Redfield, the electrical lightning of the telegraph demonstrated the uniqueness of its role in communication.

Both Shearman and Redfield and the Texas court—which heard a number of other death message cases in the wake of *So Relle*—saw the presumed sublime power of telegraphy as fundamental to telegraph companies' responsibilities to the emotions of the public. As the Texas court explained, telegraph companies had been given "special franchises and privileges under the law," in order to help them "to furnish for compensation the means of rapid communication." This "rapid communication" likewise made the telegraph an ideal means of sending emotionally urgent messages whose speedy delivery was of great importance—provided, of course, they were actually delivered with the speed the telegraph promised. Given its high-technology status, the court reasoned, "the resort to this mode of transmitting information should of itself be held sufficient notice to the company's agents that as between the sender and the party to whom sent, the message is deemed to be of some importance."[59] For similar reasons, Shearman and Redfield asserted that "as compared with almost any other kind of business, the care required by a telegrapher would be called 'great care.'"[60] It was the very sublime electrical power of telegraphy that required telegraph companies' great duty to the emotions of the public.

Even as these decisions were being offered, American conceptions of emotion were beginning to change. In a fairly short period from the 1890s to the turn of the century, the *So Relle* court's emphasis on telegraph companies' public emotional responsibilities moved from being seen as common sense to a kind of relic of a bygone era of overwrought sentimentality. The authors of an 1893 legal treatise argued that most courts agreed with the *So Relle* approach.[61] In 1905, however, an article in the *Columbia Law Review* suggested that "the weight of judicial authority is opposed to the Texas doctrine, and denies a recovery of damages for mental anguish only, resulting from negligent failure to deliver a telegraph message."[62] Evaluating the recent climate of these cases for mental anguish, a 1909 article in the *Yale Law Review* similarly asserted that "upon examining the cases it is apparent that the weight of authority is opposed to this doctrine."[63] The *So Relle* court's particular argument about the responsibilities of telegraph companies had gone the way of the Victorian style of emotional display.

The celebrations and concerns surrounding the communication of feeling made possible by the telegraph played into the complex public conception of emotion so prominent in the nineteenth century. Like the public love letter, the most intimate messages sent via telegram were subject to the public perusal of the telegraph operator, who thus could be held responsible for a portion of their emotional consequences. By "linking national hearts in one," the sublime power of the telegraph seemed to offer a technological manifestation of the public nature of emotion, uniting people in one exquisite nerve. While this shared emotional stimulation was often celebrated, as it had been by the advocates of the Great Awakening, it also began to make people uneasy, as the above concerns about the intellectual effects of electricity demonstrate. As the United States expanded its communication technologies in the early twentieth century, social researchers, clergy, educators, businesspeople, and numerous others increasingly stressed the private, individual aspects of emotion. For reasons of both social welfare and personal gain, people were encouraged more and more to keep their emotions to themselves. This reflected changing attitudes about both human feelings and the technologies that communicated them.

Technology and Emotion in a High-Speed Culture

In his own discussion of the history of American emotions, Peter Stearns has argued that the early twentieth-century United States saw the emergence of a dominant emotionology of "American cool," which stressed an individualized, controlled emotional demeanor that contrasted in significant ways with earlier American attitudes about emotion.[64] In line with the above discussions, Stearns suggests that many of the more impassioned ideas of emotional expression of the eighteenth and nineteenth centuries were eventually seen as inappropriate by early twentieth-century Americans. For instance, "fearsome animal dances associated with folk Christmas" had been part of many nineteenth-century American Christmas celebrations, and it was not until the early twentieth century that "figures like the German-American Belsnickel, a fur-wearing variant of St. Nick who carried sticks for beating bad children while recording their names in a punishment book, disappeared entirely."[65]

Working in Stearns's emotionological tradition, Linda Rosenzweig illustrates that women of the nineteenth and early twentieth centuries had expressed deep sentiments for their female friends, often using the same romantic language associated with communication between lovers. However, Rosenzweig argues, changes in psychological theories of friendship and societal norms about same-sex relationships during the first two decades of the century saw these expressions cool down, as women were supposed to save much of this romantic emotionality for their newly important "companionate" marriages.[66] Similarly, as Michael Barton has demonstrated, although stories of tragedy in early twentieth-century newspapers provided detailed, gory accounts that read like gothic horror, by the 1920s and 1930s American journalism had largely put these aside for more sterile accounts that rejected such horrifying images.[67] In these ways, Stearns and his adherents argue, much of the Victorian conception of the passions was replaced with a more controlled, cool conception of emotion.

The cultural negotiations—and contradictions—through which Americans posited this cool emotionology take vivid form when viewed in relationship to attitudes about technology in the early twentieth century. The idea that the telegraph could transmit emotion gestured toward a more electrical conception of feeling that was beginning to take shape during the nineteenth century and that reached a sort of pinnacle in the rhetoric I identify as media physicalism. If emotions could be communicated via the telegraph—here picking up some of Franklin's older sense of communication—then they must in some sense themselves be electrical. Indeed, the social scientists I discuss in chapter 4 identified emotions directly as a kind of electrical conductivity of the body. This suggested that the emotional body had some of the same sublime features as electricity itself; people's bodies were seen to vibrate with the same throbbing intensity as the national heart of the telegraph. While this extended the reach of the human senses, it also suggested that people's feelings could lead to all manners of dangerous technical effects, including overload, speed, and shock.

The sense that the new technologies were overloading or speeding up the culture was prominent in the early twentieth century. The English psychologist Graham Wallas opened his 1914 book *The Great Society* with an observation about the chaotic effects of emotional and technological connectivity:

> During the last hundred years the external conditions of civilised life have been transformed by a series of inventions which have abolished the old limits to the creation of mechanical force, the carriage of men and goods, and communication by written and spoken words. One effect of this transformation is a general change of social scale. Men find themselves working and thinking and feeling in relation to an environment, which, both in its world-wide extension and its intimate connection with all sides of human existence, is without precedent in the history of the world.[68]

The Great Society was one of speed, connectivity, and, ultimately, dangerous stimulation.

Wallas's arguments resonated strongly with his American counterparts. The journalist and cultural critic Walter Lippmann, who took a class from Wallas while he was a visiting professor at Harvard and to whom Wallas addressed the preface of *The Great Society*, similarly decried the new state of connection. For Lippmann and Wallas, a problem resulted from the fact that the scale of the community had grown so much that people could not know the others on whom their livelihood and culture depended. The Great Society was a large-scale accumulation of anonymous, chattering voices, competing with each other for public attention. Lippmann argued in his 1922 book *Public Opinion* that the size of the Great Society made it very difficult for any individual to grasp the affairs of the public as a whole. The lives of his contemporaries were filled with too much information from disparate sources and on too many different topics. As a result, Lippmann claimed, "political opinion on the scale of the Great Society requires an amount of selfless equanimity rarely attainable by any one for any length of time."[69] Given the stress of making sense of this complex of information, Lippmann wrote, "the private citizen today has come to feel rather like a deaf spectator in the back row, who ought to keep his mind on the mystery off there, but cannot quite manage to stay awake."[70] Although the American philosopher

John Dewey famously disagreed with Lippmann about the solutions to these early twentieth-century problems, he viewed the Great Society in a similarly negative light. "The Great Society created by steam and electricity may be a society, but it is not a community," he wrote. Like Wallas and Lippmann, Dewey held that "the invasion of the community by the new and relatively impersonal and mechanical modes of combined human behavior is the outstanding fact of modern life."[71]

Concerns about the scale of the Great Society created parallel concerns about emotion. For instance, at the turn of the century a range of social scientists became interested in "crowd psychology" as they attempted to understand the emotional effects of the connections made possible by this new technological age. In a defining discussion of the topic, the French writer Gustave Le Bon held that "the recent development of the newspaper press by whose agency the most contrary opinions are being brought before the attention of the crowds" contributed to the emergence of emotionally keyed-up masses that could not rationally reflect on their behavior.[72]

Although Wallas was critical of crowd psychology, many of his own positions suggested similar problems with the emotions of the Great Society. In discussing the development of advertising, Wallas argued that

> young men of good education, naturally warm feelings, and that delicate sense of the emotional effect of words which, under different circumstances, might have made them poets, are now being trained as convincing liars, as makers, that is to say, of statements, to whose truth they are indifferent, in such a form that readers shall subconsciously assume the personal sincerity of the writer.[73]

Wallas saw the subconscious power of such advertisements as akin to "telepathy"[74]—they worked subtly on the minds and emotions of their audiences, rather than through the more brute forms of suggestion and imitation he believed crowd psychologists attributed to them. Discussing the impact of news and other stories, Lippmann argued that "the account of what has happened out of sight and hearing in a place where we have never been, has not and never can have, except briefly as in a dream or fantasy, all the dimensions of reality." However, he added, "it can arouse all, and sometimes even more emotion than the reality."[75] As had been the case for some critics of the telegraph, the apparently extreme connectivity of the Great Society promised to stimulate mass emotion in some potentially threatening ways.

If notions of the technological overstimulation of emotion pervaded thinking about the Great Society, it was because so many technologies had come to be seen as powerful forms of emotional stimulation. Besides the communication technologies that are the primary focus of this book, discussions of the emotional mania associated with the automobile illustrate another fascinating case. A 1903 *New York Times* article explained that "the racing motor has given us a new disease," which writers variously termed "motor intoxication," "speed madness," and "speed mania." In giving owners and their chauffeurs the capability of travelling at significant rates of speed, the car had apparently created a drug-like new stimulation, with debilitating individual and cultural effects:

> The mental and moral states of the chauffeur [while speeding] become abnormal, the change being not unlike that by which Dr. Jekyll was transformed into Mr. Hyde. When the madness has possession of him the chauffeur becomes reckless, vindictive, furiously aggressive, and is swayed and controlled by whatever angry or insane impulse seizes him. A high rate of speed works him into the kind of nervous excitation which makes the person suffering from alcoholic stimulation indifferent to consequences, and eager only to gratify his momentary insane impulse.[76]

Intoxicated in this way, automobilists took pleasure in the very threats their manic driving created. "The automobilist is not less humane than other people," suggested another *New York Times* article, "but undoubtedly he does derive great satisfaction and excitement in seeing how closely he can miss his fellow-citizens who are riding or walking on the same thoroughfares through which he recklessly whizzes."[77]

Despite the gendered nature of these comments, men were not the only ones who seemed to fall victim to speed madness. In 1910, the writer Kate Masterson claimed that "woman—the Twentieth-century production—has sold her birthright of emancipation. She is slave to the Motor-Car—the great, luridly painted, furious, rankly odorous machine that now whizzes through the streets of every great city in the world." The whizzing of the car impacted "not only the athletic, sport-loving feminines." Rather, "even the physically delicate, [had] learned to glory in the dangers of the automobile and to laugh at narrow escapes from smash-ups, as if the game were as simple as croquet." In Masterson's analysis, the delicacy of the cult of true womanhood had been all but destroyed by the car's arrival. "Mercy and kindness have been feminine traits allowed to even the least of them," she argued. "But all of these loveable womanisms have been lost in the new delirium, the mysterious intoxication of the devil wagon." The whizzing automobile not only made the streets more dangerous, but threatened the basic fabric of feminine society. For Masterson, the car had "[made] the quiet joys of home and hearth seem only amusements for the peasant. It [had] brutalized every fine feeling—with all its tremendous rush of power and speed and consequence!"[78]

The idea that driving a car could be dangerously intoxicating or could create a kind of addictive mania was an important implication of the new, technologically focused conception of the emotions (though it had deeper roots in Socrates's conception of speech as a *pharmakon* as well). Like the electrical vibrations of a telegraph wire, emotions were electro-physiological forces that vibrated in response to the clicks of the telegraph or the acceleration of an automobile. In this sense, emotions were powerful electrical forces *within* people's bodies, but they were also somehow alien to them—coming from outside through various sorts of external, technical stimulations. From here, it was an easy step to imagine emotions as a type of drug. They entered the body and seemed to overtake it with a kind of frenzied enthusiasm. Here was an extreme version of the private conception of emotion that had displaced the Victorian idea of the passions at the turn of the century. Emotions were not so much a social good to be shared as an eminently individual problem to be personally channeled, controlled, or otherwise overcome.

A 1908 article in a journal of phrenology—the science of reading people's personalities through the bumps on their skulls—made the ubiquity of speed mania and its individual, bodily consequences very explicit. "Speed mania is a term well applied to American life in general," the author claimed. "To-day this man has been killed in an auto accident. To-morrow that Wall Street man falls dead at his desk. In both cases speed mania was the cause." In laying out the phrenological implications of this, the author explained that "speed madness, like other forms of mental derangement, can arise from a number of entirely different causes." Insufficiently large faculties of cautiousness, vitativeness (an "instinctive love for life"), and conscientiousness, all of which were represented in the phrenological characteristics of an individual, could make one vulnerable to speed mania. This vulnerability was exacerbated by excessively large faculties of combativeness, bravado, and sublimity. "Sublimity is in itself a great faculty," the author argued, because it "enables humanity to appreciate the vast, grand, endless, sublime, magnificent, wild, terrific; the lightning's vivid flash, the rolling of thunder, the commotion of the elements, etc." However, he quickly added, "this faculty is also a cause of speed madness, as it enjoys anything that is extreme and uncommon. An autoist with larger sublimity will exult in flying at lightning speed, and if there is neither large cautiousness nor Vitativeness to balance Sublimity, he will even exult in and like the sense of danger."[79] The sublime impact of the new technologies was written on the bodies of the automobilists and others enthralled in the new madness for speed.

Figure 1.2
Phrenological image depicting someone suffering from "small cautiousness."
Reprinted from E. Favary, "The Evolution of the Automobile," *Phrenological Journal and Science of Health* 121, no. 1 (1908).

The supposed stimulation of the car—captured in speed mania—thus reflected a particular sense of technological sublimity that seemed to demand a controlled emotional demeanor. In addition to the physical danger posed by a speeding car, its intoxicating powers to turn women away from their feminine sensibilities, to turn a chauffeur into a Mr. Hyde, and to make everyone who succumbed to its powers indifferent to the sufferings of their fellow travelers pointed to the cultural problems with emotional overstimulation. At the same time, the fact that cars had such narcotizing power suggested that drivers needed to exercise a strong degree of emotional control in order to operate them safely. In a discussion of ongoing automobile accidents, a *New York Times* article stressed that, in moments of disaster, "the machine is liable to become unmanageable, either from some mechanical defect, or from an access of nervousness on the part of the driver which leads him to do, in the moment of peril, precisely the wrong thing."[80]

Elaborating on how "nervousness" might lead to auto accidents, another newspaper article asked readers,

> Will you have the presence of mind to put on the brakes and pull over to one side to avoid an accident? Or will you be frozen to inaction for the brief second that will be your time margin for safety and apply the brakes too late? If your wife is driving, will she scream and put her hands to her face, or will she do the right thing?

The article recounted tests of automobile driving performed at George Washington University, which reportedly found that men were better drivers than women. "A woman is naturally more emotional than a man," the article explained. "A 'close call,' which necessitates extremely quick action on the driver's part, often completely unnerves a woman for the rest of the trip. A man's greater physical strength and greater familiarity with machinery, as well as his better emotional control, make him a safer driver than a woman."[81]

As these comments suggest, the assumed seductiveness of the automobile brought out a range of other concerns, including concerns over who was and was not equipped with the emotional self-control necessary for the speed and intensity of the new century. The same belief in the hyperemotional nature of women that supposedly made them bad drivers was also a primary obstacle against universal women's suffrage. According to popular wisdom, women were too emotional for either the car or the ballot box, and the unseemly excitement of both the roadway and politics threatened to destroy what was left of their delicate female sensibility. Concerns about speed mania were expressions of a range of other anxieties concerning the forces of modern American life, including the increasing entry of women into the public sphere.

A *Los Angeles Times* article describing a town meeting in Brooklyn Heights, California, in which the dangers of speeding trains were discussed addressed another group with dangerous emotions. The meeting had apparently been "disgraced" by "blatant socialists," who engaged in "the turgid oratory of too fervid agitators," including "vituperation, thickly larded blasphemy, ribaldry, and expressions of violence." Still more dangerous, according to the article, was a statement in the *Evening Express*, a newspaper owned by Edwin Earl, that suggested that the speeding of train cars resulted from the train company's efforts to make as much money as possible during the course of the day. The *Los Angeles Times* writer refuted this, suggesting that speed mania was rather the symptom of the desire for "getting there" identified in so many discussions of the public's latest madness. In attributing this need for speed to the company's financial motives, the *Express* had "become the mere tail to Hearst's socialistic kite" and was "appealing by every means it can think of to the passions of the lowest and most dangerous elements of the population. Hence Millionaire Earl's daily output of demagogic rant against the capitalists."[82]

Between the speeding cars, ranting socialists, and seductive advertisements, this period seemed for many to be a flurry of stimulation, giving edicts of emotional control a particularly powerful warrant. A strong faculty of emotional control would presumably offer a defense against the intoxicating powers of the automobile, socialist, and adman alike, and was assumed necessary for good health more generally. In a newspaper column syndicated throughout the 1920s and 1930s, Dr. William Augustus Evans suggested emotional control as a preventative for everything from dyspepsia to insanity. In response to a letter from a woman worried about her eighteen-month-old daughter's temper, Dr. Evans wrote that "the remedy is character training. Unless she is trained in emotional control she is in trouble."[83] Similar advice was offered by a range of scientists, doctors, teachers, and other advocates of a slower, more restrained culture.

At the same time, a variety of manufacturers and media producers explicitly championed the emotional stimulations brought about by their products. In the face of complaints about speed madness, an Oldsmobile advertisement from 1905 celebrated the company's cars as "a public utility for every avenue of business hurry. Always ready when you are—a race horse when you want speed." A Rambler ad from the same year claimed that "no other machine at the same price has the beauty, speed and reliability of our Surrey Type One," and Ford boasted that its Model B had "more power for its weight than any Automobile of similar type."[84] These companies were selling the very speed that a range of critics found debilitating.

Despite such apparent disagreements, these competing messages ultimately played into some very similar ideas about emotion and technology. Both speed critics and car companies imagined that a particular emotional power was located in the technology of the car itself. The car was a "devil wagon" because of its intoxicating power to overwhelm its passengers with a mania for speed. Ford, Oldsmobile, and Rambler depended on a similar assumption. They tried to persuade consumers that their cars would provide access to a set of emotional stimulations unavailable without them. Horsepower ratings and phrases such as "double opposed motor" were intended to signal the emotional experience awaiting the driver. Both critics of speed and its advocates offered competing versions of emotional control as well. Speed critics explicitly condemned emotional excess; they reacted against what they imagined as a stream of speed-intoxicated drivers selfishly hurtling down the roads. Those companies manufacturing, marketing, and selling the speeding autos offered another way individual consumers could take control of their emotions. In purchasing a properly equipped car, automobile ads promised, consumers would guarantee themselves a particular set of emotional experiences. Americans of the car culture thus had at least two ways to manage their emotions. They could either resist the seductive automobile completely, or they could purchase one that stimulated them in a sufficiently powerful, appropriate manner.

In making their case for buying a car, auto manufacturers linked the physical sensation of whizzing down the road with a range of other feelings. Thorstein Veblen had identified what he termed "conspicuous consumption" in 1899, and advertisements stressing the brand identities of various products—as opposed to simply describing their characteristics—had become prominent only at the end of the nineteenth century.[85] In tandem with the rise of the automobile, a range of products were being marketed for the various physical and cultural benefits they presumably offered consumers. A Coca-Cola advertisement from 1905 suggested that the drink could "invigorate the fatigued body and quicken the tired brain." It would also apparently bring an elevated sense of cultural standing. The ad depicted four apparently well-to-do people, sitting in an automobile, receiving their Cokes from a waiter's silver tray.

Car manufacturers promised a similar combination of physical experience and elevated cultural sensibility. Hastily whirring past the lowly pedestrians, the automobilists could flaunt their personal enjoyment through the public streets. But rather than simply celebrating bold, public expressions of passion, as had been done in conjunction with the Great Awakenings and other eighteenth- and nineteenth-century cultural movements, the marketing of the auto and these other products endorsed an overwhelmingly private, controlled sense of feeling. As one's emotional experiences became located in automobiles and other consumer goods, one's feeling life became a matter of assembling the proper collection of emotional stimulants. Each individual was to manage his or her emotions through the careful selection and use of the appropriate products.

Warring Emotions and the Associational State

Early twentieth-century auto manufacturers, journalists, medical professionals, phrenologists, and social critics offered a conflicted yet mutually reinforcing rhetoric that emphasized the powerful emotional stimulations of the new technological age, focusing in particular on its supposed chaos and speed. Given this emphasis, it should not be surprising that American concerns about technology and emotion reached a heightened attention around World War I. The war seemed to offer an extreme example of the emotional stresses that pervaded the culture as a whole. Like discussions of "speed mania," arguments about the emotional impacts of war framed technology as both a problem and a solution and created rhetorical bedfellows of thinkers who seemed, on their surface, to be opposed to one another.

Linking technology and emotion in vivid fashion, World War I soldiers made heavy use of a late nineteenth-century invention that *Popular*

Science Monthly called "the world's greatest terror." The machine gun had been created in 1883 by Hiram Maxim. "When smokeless gunpowder was invented," the *Popular Science Monthly* article explains, "the Maxim machine gun came fully into its own; for it was possible to take up a concealed position and squirt death without betraying clouds of smoke."[86] Similarly, a newspaper story in the *Washington Post* highlighted the sublime terror of the "man-made sharks of steel" and "man-made vultures of air" that were combat submarines and airplanes, the latter "striving to peck out the eyes of the man made monsters beneath the waves."[87]

If the high-speed, advanced technological nature of automobiles raised concerns about emotional stimulation, writers were understandably worried about the emotional impact of these new war technologies. "Shell shock" became a frequent topic in both British and American newspapers. Symptoms associated with shell shock included paralysis, deafness, multiple personality disorder, and other psychic and physical maladies. An article in the *Los Angeles Times* explained that "shell shock disables a man without putting upon him any wound or visible mark of external injury," adding that, according to medical experts, the condition was "singular and particular to the intense artillery war." Another *Washington Post* article held that armies were "employing as engines of terror and destruction in the present war bombs laden with poisonous gases and monster shells which drive soldiers deaf, dumb, blind, and insane, and deprive them of taste and smell."[88] The new technologies of the war were assumed to be taking a massive psychological toll on troops.

Soldiers were expected to master the same technologies that subjected them to this emotional and physical trauma. They needed to be proficient with artillery and machine guns, and needed to know a variety of technical details about the metal "sharks" and "vultures" they engaged—or piloted—in the field. In fact, soldiers were to be knowledgeable about technologies that stretched beyond the battlefield itself. As part of a series of tests given to assess and classify soldiers based on their intelligence, candidates given a picture of a phonograph were supposed to identify that it had a missing "horn."[89] In addition to these technologically literate soldiers, the war also made use of "Hello Girls," who served as telephone operators for the war effort, making vital connections between troops in Europe as well as between Europe and America. By the time the war ended, these operators were stationed at nearly every major telephone exchange in France.[90] According to one naval magazine, "every minute of the time that these girls sit before the switchboard, their fingers are nimbly at work. Almost every second a little light pops up and demands attention. It's a tedious position, and the biggest essential is self-control."[91] American soldiers and "Hello Girls" alike were expected to develop a mastery over both specific technologies and their own emotions.

George Crile's 1915 book *A Mechanistic View of War and Peace* offers an interesting example of how World War I fit with and impacted the larger understandings of technology and emotion that were prominent during this time period.[92] Crile was a successful American surgeon. He cofounded the Cleveland Clinic, was selected as an honorary fellow of the Royal Surgeons of England, and had, in 1906, performed the first blood transfusion in the United States.[93] Among his many interests, Crile held a great fascination with the physiology of the emotions. While serving as a medical officer in France, he collected internal organs of recently killed soldiers in order to study their levels of excitement. On hunting trips to Africa, he likewise studied the adrenaline levels of wild animals he had killed.[94] As a surgeon, Crile was especially interested in questions of surgical shock. In a series of experiments, he performed surgeries on dogs as he monitored their respiration, blood pressure, and other physiological responses. His goal was to document the kind of trauma or shock that was experienced during surgery.[95]

Crile's studies of shock set the groundwork for his views on the emotional impacts of war. Like popular commentaries on machine guns, Crile's *Mechanistic View of War* emphasized the emotional impact of artillery fire, which, he argued, "shakes the body, and often produces a molecular change in nervous tissue." Artillery fire also depersonalized battle as compared to fighting with a rifle or bayonet. As Crile put it, the "artilleryman . . . has no personal contact with the enemy, but suddenly finds himself under a scorching fire, from a source which he cannot ascertain, from an enemy he cannot see." For these reasons, Crile suggested that artillery warfare was "like quarreling by telegraph,"[96] reiterating in an extreme form some of the nineteenth-century anxieties about waiting on the "tenter-hooks" for a distant, emotionally intense communication. From Crile's perspective, warring soldiers were simultaneously hyperemotionally engaged and emotionally disconnected.

For Crile, this conflicted emotional overstimulation was not unique to soldiers. The declarations of war had stimulated the nerves of the citizens in the various warring countries. Once these countries entered into the war, "the *kinetic system* of each individual was activated." As a result, Crile explained, "there was an increased output of adrenalin, of thyreoiodin, of glycogen; and an increased mobilization of the Nissl substance in the brain-cells, from all of which there resulted an increased transformation of energy in the form of heat, motion, or chemical action."[97] In fact, because soldiers engaged in such activities as marching and singing, which helped to dissipate their increased energy, Crile held that "the kinetic systems of the soldier during mobilization are less strained than are the kinetic systems of those he left behind."[98] Crile believed that the emotional strain of war was in many ways more powerful for those who remained at home.

In Crile's rhetoric, the emotional stimulation of wartime was a reflection of larger issues that pervaded the modern world more generally. Like other thinkers of the period, Crile stressed the adverse effects of the era's presumably heightened emotionality. In the era of the machine, Crile asserted, all life had become like a war, as average people were engaged in an ongoing emotional, physical, and technological struggle:

> These descendants of the cave man have captured and domesticated lightning; they have enslaved the world with a copper nervous system which enables them to activate the action patterns of, and in turn be activated by, hundreds of millions of the human race. A slight change in the chemistry of a human brain cell may wreck a bank in India, fire the first gun in a great war, or break a woman's heart. Such is the web of life man has woven and by means of which he so completely dominates the earth.[99]

This complex, technological web of life created an "excessive motor and emotional driving" that destroyed the body through disease and the psyche through constant stress.[100] It was as if all modern citizens were living through their own continuous cases of shell shock.

Like so much of the wider rhetorical ecology of the period, Crile saw both war and emotional overstimulation as the symptoms of the technological nature of the modern world. Nonetheless, in Crile's analysis these issues required a *technological solution*. This was because humans

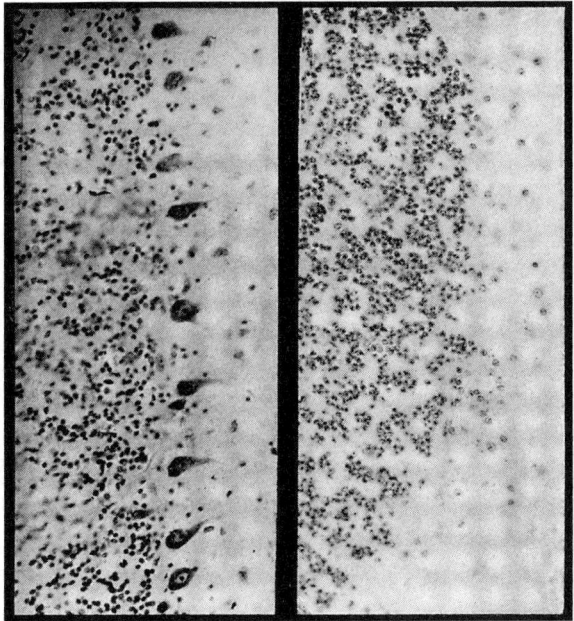

Figure 1.3

Two images of the cerebellum showing the effect of extreme fright. The image
of the left depicts a normal cerebellum and the image on the right depicts a
cerebellum after the frightful experience. Reprinted from George Washington
Crile, *A Mechanistic View of War and Peace* (New York: Macmillan, 1915), 81.

themselves were essentially machines. "Man is a mechanism that acts as
a machine, that is a machine like a locomotive or an automobile," wrote
Crile.[101] Imagining the human brain as a kind of media technology, Crile
asserted that it was similar to both "the apparatus of a wireless receiving station" and a "moving-picture film running from birth to death."[102]
Building from this view, Crile argued that people performed in the same
predetermined ways as other technological systems. "The reactions of the
human mechanism," wrote Crile, were "as inevitable and as true as [were]
the reactions of a man-made machine." "A wheelbarrow," he asserted,
"cannot perform the work of an automobile, but the difference between
the wheelbarrow and the automobile is less than the difference between
the cannibal and the scholar."[103] Human beings were but reflections of the

era's most sublime technologies, and people's relative health depended upon how well they played their presumably high-tech, mechanistic role.

Owing to this thoroughly mechanistic view of human being, Crile asserted that the key to managing humans' emotions and ending war—if it could be ended—was creating new designs for human beings them-selves. This meant intervening from birth in the "moving picture" of a person's life. Drawing connections between industrial manufacturing and human development, Crile held that "the newborn infant is only the plastic clay from which the real man is created."[104] As a way of dealing with the bewildering amount of information and stimulation available to humans in the early twentieth century, Walter Lippmann had proposed that a class of social scientific experts be enlisted to make sense of impor-tant facts for the broader public. Given his mechanistic focus, Crile went a step further. He suggested that a class of "supermen" could develop sys-tems for better molding and designing the young human "mechanism":

> If the human animal were under the domination of beings as superior
> to him as man is superior to the domestic animals, we might expect that
> education would be exploited as efficiently as war has been exploited and
> that there might be built up a civilization freed to some extent from its
> menacing phylogeny.[105]

If modern life was like a war, Crile reasoned, then it should be managed like one. This meant training ordinary citizens in the same emotional and technological control employed by the finest soldiers and "Hello Girls," and thus crafting human mechanisms that could match up to the sublime technologies with which they interacted every day.

Crile's appeal to "supermen" experts and his mechanistic view of social change reflected several larger conceptual ideas that devel-oped around the war and continued in the years that followed. First, it reflected ideas about what Ellis W. Hawley has called the "associa-tional state." In the early twentieth century, business professionals were increasingly assumed to be among the best advisors of government administrations. As Hawley explains, "the watchwords of the progres-sive era were *organization* and *professionalism*, and by many Americans the period's business oriented 'organizers' and 'professionalizers' were perceived as the new agents of social progress."[106] The associational perspective put an emphasis on self-control and self-regulation. Trade unions and other professional groups—rather than government officials—would control various practices internally. Herbert Hoover, first as the secretary of commerce and then as president, was one of the biggest advocates of this associational vision, believing that cooperation among various professional organizations would create a superior form of government controlled overwhelmingly by private concerns.[107]

Among those professionals that were seen as central to the associational state, engineers played an especially important role. Although he spoke out against the hyperconsumption of the new age, Thorstein Veblen championed the engineer as an important figure in the industrial system. Business interests had corrupted the industrial technologies of the new century by slowing down production to maintain the most economically profitable supply. This amounted to vast "waste" and "inefficiency." If the industrial system could be run by technologically focused engineers—"the keepers of the material welfare"—the system could be run more efficiently, delivering on its more utopian promises of material abundance for everyone.[108]

In seeing efficient use of technology as a means to a more progressive—even anticapitalist—world, Veblen was helping to inaugurate a viewpoint that would become an important lens for understanding early twentieth-century technology. "Technocracy," as the engineer William Henry Smyth and others called it, dealt with "the organizing, coordinating and directing through industrial management on a nationwide scale of the scientific knowledge and practical skill of all the people who could contribute to the accomplishment of a great national purpose."[109] Like Veblen and Crile, Smyth and the other technocrats viewed the world through a largely mechanistic perspective. Life had become so thoroughly mechanized that any serious social change would need to be carried out by engineers, mechanics, and other technological experts who could be sure that the machines served the people at large. Otherwise, everyday citizens risked an autonomic existence that simply reinforced the economic power of "robber barons" who kept these technologies under their control.

Notwithstanding the obvious ideological disagreements between the associationalists and the technocrats—illustrated perhaps most clearly in the conflict between the outlook of the Republican Hoover and that of the anticapitalists Veblen and Smyth—these two positions held much in common. Both saw presumably impartial experts as the best guides for the government and society at large. Likewise, both put a heavy emphasis on the efficient use of technology. Before his presidency, Hoover—"the Great Engineer"—led a study entitled *Waste in Industry*, in which he and a series of other experts made suggestions for improving the mechanisms of a range of American industries.[110] It was a testament to the period's conflicted ideas about technology that two presumably politically opposed movements could have such similar perspectives as far as the technology question was concerned.

It should not be surprising that both associationalists and technocrats were especially vocal in the immediate postwar period, as well as in the period following the Depression. Like the war, the Depression highlighted both the sins and virtues of technology and technological ways of thinking. Public interest in the technocracy movement peaked in the 1930s as the nation sought ways to deal with its economic crisis. Echoing Veblen and Smyth, Howard Scott, another engineer and a central figure in the movement, argued that the Depression had been heightened by the mismanagement of industrial machinery. Despite the nation's vast resources, Scott wrote in 1932, "we have, nevertheless, failed to profit from technological advances, and accordingly find ourselves, for the first time in history, with an economy of plenty existing in the midst of a hodgepodge of debt and unemployment."[111] Among other things, Scott argued that increased mechanization had put many people out of work, as in many industries a single skilled machine operator could do the work that had previously taken numerous people.

Scott and other technocrats of the 1930s were accused of spreading gloom and of seeing machines as the ruin of humanity; but, again like Veblen and Smyth, Scott insisted on the utopian possibilities of a properly managed technological system. "Technocracy points out that this continent has no cause for gloom, or fear of chaos, but that we must face the inconvenience of change—that this continent today stands on the threshold of a new era of wellbeing," he explained. The very technologies that were displacing workers could also produce an abundance that made the price system and thus traditional labor obsolete. "The high road to this new era," he reiterated, "can be one of orderly progression under technological control."[112] In a like manner, and in spite of taking much of the blame for the failing economy, many of the ideas associated with Hoover's "associational state" carried over into the New Deal, especially its emphasis on experts and "technological imperatives."[113] Technological progress remained a popular solution for presumed technological disorder.

In addition to their conflicted interests in technology, the associationalists and the technocrats also shared a concern for emotional control. Among the other problems identified in Hoover's *Waste in Industry*, "fear and nervousness" were seen as detrimental to an efficiently working industrial system. Reflecting general concerns about how heightened technologies increased emotional stress, the study suggested that the more skilled workers—those who worked most closely with the most advanced technologies—experienced a high degree of demand on their central nervous system.[114] William Smyth blamed resistance to the progressive ideas of technocracy on an overwhelming irrationality and fear that were products of humans' "saurian primordial ancestry," which needed to be overcome if the nation was to take advantage of scientific and technological ways of thinking.[115] For both associationalists and technocrats, mastering technology meant replacing one's own emotions with a more machine-like rationality.

The concerns about technology and emotion that surrounded World War I and the Depression and that were illustrated in both associationalism and technocracy were metonymic of the wider apprehensions that characterized much of the early twentieth century. Advancing technologies, the arguments went, created additional emotional stimulation. Nineteenth-century Americans had celebrated a range of powerful emotional expressions; responding to the supposedly heightened emotionality of the new century, they increasingly stressed the emotionally controlled demeanor of "American cool." Far from a rejection of technology, this amounted rather to a wide-scale adoption of a technological ideology—the mechanistic perspective through which Crile believed he could solve a number of the nation's problems. Among other things, this entailed seeing people themselves as machines. If people operated with some of the same cool precision as automobiles, airplanes, and telephones, then they presumably had a chance of escaping the destructive emotional stimulation that accompanied those technologies.

Administrative Research and Technological and Emotional Control

The communications media of the early twentieth century offered a special instance of these tensions about emotional stimulation, as the following chapters illustrate. If automobiles impacted people's bodies in a phrenological manner, then communication technologies were certainly seen to be impacting individual physiology and speeding up the culture. For both good and ill, the radio gave people rapid access to a host of feelings that would otherwise be out of their reach. Whereas the car stimulated emotion by moving people through space, however, the new communication media, like the telegraph before them, seemed to transport emotions themselves. This gave these media both added promise and peril, extending the national heart to still wider areas of the culture. When people switched on their radios, they would be swept into a whizzing of feelings that seemed to transcend time and space entirely. Likewise, just as the car promised not merely the sensations of literal movement, but movement up the social ladder, so the new media promised their own elevation of class and cultural standing. In both celebrations and concerns about the power of the new media, a range of thinkers supported the scientific superiority of modern technology. A technologically advanced radio—like a well-equipped motor car—would demonstrate the owner's higher position along the evolutionary scale of modernity. However, the more developed these technologies became, the more powerfully they were believed to transmit and stimulate the emotions. It was the very hyperemotional stimulations of these technologies that demonstrated their modern, scientific, rational superiority.

In these ways, beliefs in the sublime power of communication technology at the beginning of the twentieth century created an interesting set of dilemmas, suggesting that emotional overstimulation was a necessary result of the progress of scientific and technological advancement. The range of ways that communication technologies were both celebrated and condemned illustrates the complicated ways that this paradox played out. These struggles also demonstrate the various anxieties about modernization for which debates about technology and emotion became a proxy. Worries about cultural speed, socialism, women's liberation, World War I, and the Great Depression created a variety of other individual and national tensions. Amid this complex of cultural changes, people worried about their own and others' race, class, gender, and connection to the nation. In questioning the capacity of radio and other media to transmit people's emotions, Americans of the early twentieth century were concerned with their own abilities to feel and connect in what seemed to them a complex, fast-paced, rapidly changing culture.

Communication technologies codified these concerns in other ways as well. In debates about how radio and similar technologies could enhance or hinder the emotional connections of the public, the associational and technocratic concerns of the early twentieth century found a powerful expression. As the next chapters illustrate, a wide range of experts focused their concern on the new recording and broadcast technologies, creating strong relationships between academic media research and professional media production. This burgeoning "administrative approach" to communication drew together associational and technocratic thinking. The new media researchers were critiquing these developing technologies even as they employed them in their research. Given the centrality of these approaches to media research, although on a broad scale associationalism and technocracy may have largely died in the 1930s—as Hawley and others have argued—they would become an intrinsic part of the country's understanding of media. The general approach to media that became ensconced in the Communications Act of 1934 assumed both the basic altruism, or at least inevitability, of corporate media interests and the sublime power of media technologies.

Likewise, the larger emotionology of American cool was influenced by concerns about technology and also became a central factor in how early twentieth-century media researchers, and the broader culture, made sense of the new media world. The presumed emotional power of stereoscopes, phonographs, the radio, and motion pictures promised great rewards to the culture even as they threatened it with the growing shock of modernity. It was against this paradoxical backdrop that the burgeoning media physicalism imagined its ideal citizen. Because emotions were essentially technological, to control a technology and to control one's own emotions were one and the same. Everyday people, entrepreneurs, and social scientists alike were to master these technologies—and themselves—by adopting a veneer of mechanistic rationality that would become emblematic of a new high-tech, cultured citizenship.

2

Touching Images

Stereoscopy, Technocracy, and Popular Photographic Physicalism

In a mid-nineteenth-century essay, the English writer Lady Elizabeth Eastlake celebrated photography's ability to unite the populace and stimulate public sentiment:

> Where not half a generation ago the existence of such a vocation was not dreamt of, tens of thousands (especially if we reckon the purveyors of photographic materials) are now following a new business, practising a new pleasure, speaking a new language, and bound together by a new sympathy. For it is one of the pleasant characteristics of this pursuit that it unites men of the most diverse lives, habits, and stations, so that whoever enters its ranks finds himself in a kind of republic, where it needs apparently but to be a photographer to be a brother. The world was believed to have grown sober and matter-of-fact, but the light of photography has revealed an unsuspected source of enthusiasm.[1]

In Eastlake's view, the photograph amounted to "a new form of communication between man and man—neither letter, message or picture—which happily fills up the space between them." The superiority of photography grew from its ability to produce "pictures of life insurmountable in pathetic truth."[2]

The emotional power of photography was a common theme during the nineteenth century. On the one hand, many worried about the impact of this new mass art, which threatened "conflagration and anarchy, an incendiary leveling of the existing cultural order."[3] At the same time, Eastlake and others celebrated how photography made emotional imagery available to a wider population than was reached by traditional fine art, anticipating an argument that Walter Benjamin would make about motion pictures in the 1930s.[4] In that it allowed the pictures of important people, natural wonders, and other powerful images to be shared by great numbers, the new photographic technology had aided in creating, according to one headline, "science for the people."[5] To these observers, photography did more than simply communicate emotional scenes. As one writer put it, the camera had been "taught to reflect the soul's secret emotion, and to vividly portray, not the mere outlines of form and feature, but the subject as he is." In doing so, a photograph introduced the subject "for the first time to an intimate acquaintance with his own lineaments."[6] Through the poetic science of photography, people presumably experienced their own emotions—as well as the emotional meanings of natural and other objects—in a more pure light.

Photography's supposedly intimate view of emotion played an important role in nineteenth-century science as well. In *The Expression of the Emotions in Man and Animals*, Charles Darwin used photographs to both capture and convey emotion. Applying electricity to people's facial muscles, he would get them to assume a range of emotional expressions that he recorded photographically. He would then show these photographs to other people and ask them to describe the emotion the photographed person appeared to be experiencing. One of the book's images depicts "a young lady who is supposed to be ripping up the photograph of a despised lover."[7] Darwin's inclusion of this image, which he said demonstrated an expression of "contempt," highlighted the connections between emotion and photography that formed the basis of his study; the photographed woman's apparent desire to destroy the image suggested the powerful emotions that lingered in a photographic representation.

Following on much of Darwin's research, Francis Galton used photography as a tool for researching people's emotions and broader personality traits. The founder of "eugenics," which sought to classify various genetic types as a way to control criminal and other unwanted behavior, Galton developed a photographic technique that he called "composite portraiture." Galton would superimpose a series of photographs on top of each other to produce a photographic "composite" that, he claimed, "represents no man in particular, but portrays an imaginary figure possessing the average features of any given group of men."[8] Discussing a series of composites made from photographs of criminals convicted of violent crimes, Galton observed that "the special villainous irregularities in the latter have disappeared, and the common humanity that underlies them has prevailed. They represent, not the criminal, but the man who is liable to fall into crime."[9] For scientists and popular advocates alike, photographic technology saw a deeper emotional truth that was invisible to the nontechnological eye.

This chapter explores connections between ideas about emotion and the stereoscope—a binocular viewing apparatus through which two nearly identical, side-by-side photographs produce a single "three-dimensional" image. The stereoscope experienced wide popularity in the mid-nineteenth century, becoming a popular form of parlor entertainment and finding a place in the research of Galton and other scientists. Like photography more generally, the stereoscope was met with conflicted celebrations and denunciations, both of which were magnified by the presumably high-tech nature of its more realistic three-dimensional image. Although the popularity of the stereoscope—and thus arguments over its emotional power—waned in the late nineteenth century, stereoscopy experienced a dramatic renaissance in the early twentieth-century United States. During this period, companies such as Underwood and Underwood, of Ottawa, Kansas, and the Keystone View Company, of Meadville, Pennsylvania, linked the stereoscope to education and notions of national citizenship, suggesting that it could help teach subjects such as geography as well as aid in the appreciation and spread of American civilization. Both companies developed a host of practices—such as publishing books intended to accompany their stereoscope collections—that, along with their photographs, connected the emotional power of stereoscopic technology to intellectual, cultural, and moral development.[10]

Both those who celebrated the stereoscope and those who denounced it did so on the basis of its presumed technological power over the emotional experiences of viewers. Still more so than the car or even the telegraph, the stereoscope suggested the fragileness of human sensation and emotion as far as technology was concerned. It utilized a technological trick—an optical illusion—to induce an inner experience of three-dimensionality. Among those who saw the potential—and danger—of this illusion was Ernst Mach, a physicist and philosopher at the University of

Vienna who had a profound influence on the Vienna Circle, in which physicalism proper developed. In a lecture entitled "Why Man Has Two Eyes," Mach suggested that stereoscopy's power went beyond simply making things appear realistic. It could, he argued, "visualize things for us which we never see with equal clearness in real objects." Using "ghost images" and stereoscopic techniques, he suggested, one could construct multidimensional, transparent images of the workings of machines, and claimed to have already done so with anatomical images. He added that "photography is making stupendous advances, and there is great danger that in time some malicious artist will photograph his innocent patrons with internal views of their most secret thoughts and emotions."[11]

Mach's last comment could not have offered a clearer statement of the rhetoric of media physicalism as it was articulated by early twentieth-century stereoscope companies, in consonance with a wide range of academics and other thinkers on which they drew. As Keystone View Company as well as Underwood and Underwood sought to mass-produce and sell stereoscopic "tours of the world," they also sold a highly packaged vision of technological emotions. From their standpoint, the technological power of stereoscopy opened up both photographic objects and viewers themselves to particular kinds of emotional communications. Stereoscopy made possible the spiritual, aesthetical, and cultural experiences of travel not because viewers were literally transported to the scenes they viewed in their stereoscope, but because the emotions of both the scene and the viewer were brought into a kind of technological harmony. As one Underwood and Underwood publication put it, when we use a stereoscope, "our feelings are, our experience is, not that we are in the presence of a telephone, which gives out certain articulate sounds, but in the presence of a human soul."[12]

This vision of soul melding got at the highly technological view of emotions that came to dominate the early twentieth century, in which emotions gradually came to be seen as the property of technologies such as the stereoscope, rather than of people or the culture more generally (that notion of the passions that had still been active in the nineteenth century). In advancing this view of emotion, stereoscope companies also offered a fantasy of emotional control, in which viewers could take hold of their emotional lives by buying and properly employing a stereoscope. Finally, as do the other examples of media physicalism throughout this book, the rhetoric of these stereoscope companies depended on a range of class and racial distinctions that sought to present this emotional and technological control as particularly white and middle-class. The photographic physicalism of popular stereoscopy promised users a technological means of advancing both their emotional development and their social standing—at the expense of a range of assumed low-tech, emotionally undeveloped others.

Stereoscopic Realities

Experimenting with optics in the 1830s and 1840s, the English scientist Charles Wheatstone developed the earliest stereoscopic technology, which was then further developed by the Scotsman Sir David Brewster. Building on the idea that a person's two eyes tend to see slightly different images, stereoscopy utilized this visual disparity to create the optical illusion of depth. When viewed through a stereoscope, a stereoscopic photo (or stereograph) appeared to take three-dimensional form. These stereoscopic technologies soon became mass-produced, and by 1856 the London Stereoscopic Company had sold as many as five hundred thousand stereoscopes and offered as many as ten thousand stereoscope pictures from which to choose.[13] In 1854 the Langenheim Brothers American Stereoscopic Company brought stereo-viewing cards to the United States, selling stereoscopic images of scenic U.S. landscapes.[14] The popularity of stereoscopes in the United States increased still further after Oliver Wendell Holmes invented a more lightweight, portable stereoscope, which he described in an 1859 article for *Atlantic Monthly*.[15] The stereoscope remained popular through the 1860s, but dropped in popularity in the 1870s when the *carte-de-visite*, "a full-length portrait 2¼ x 3½ in. mounted on a card 2½ in. x 4 in.," became the popular photographic piece of the time.[16]

From its origin, the stereoscope became part of a complex rhetoric of technological sublimity, building on the complicated ideas that already surrounded photography. This can be partly illustrated by Sir David Brewster's understanding of science, which itself became part of the rhetoric that sustained the stereoscope. On the one hand, Brewster used science to refute a number of supernatural claims made by spiritualists and others of his time. His book *Letters on Natural Magic* attempted to explain a whole range of optical, aural, and mechanical illusions that had been or were being used to deceive the public. The speaking head of the oracle at Lesbos, the vocal statue of Memnon, and a chess-playing automaton designed by Wolfgang von Kempelen "were all deceptions derived from science, and from a diligent observation of the phenomena of nature."[17] Brewster's goal was to explain the science that allowed these deceptions to take place, and thus to rid them of their potentially dangerous power.

In one instance, Brewster was unable to offer an explanation of a session of "spirit rapping," a séance in which spirits communicate by knocking on the bottom of a table. This prompted the person who performed the séance, Daniel Home, to publicize Brewster's new belief in spiritualism. In a letter of response, Brewster wrote that although he could not explain several mechanical effects that took place during the event, he "never thought of ascribing them to spirits stalking underneath the drapery." He ended by arguing that when Home suggested that he had the power to bring the people present "into physical contact with their dearest relatives, and of revealing the secrets of the grave, he insult[ed] religion and common sense and tamper[ed] with the most sacred feelings of his victims."[18] Brewster's scientific view left no space for the mystical claims of spiritualism.

At the same time, for Brewster, there was something especially magical about science itself. In *Letters on Natural Magic*, Brewster wrote that "modern science may be regarded as one vast miracle, whether we view it in relation to the Almighty Being, by whom its objects and its laws were formed, or to the feeble intellect of man, by which its depths have been sounded, and its mysteries explored."[19] These miracles of science could be witnessed in the technological sophistication of the stereoscope. Although Brewster acknowledged that the appearance of solidity in stereoscopic vision was an optical illusion, he nonetheless championed the stereoscope's technological and emotional power. A typical photograph of a sculpture missed much of its luster and form. However, Brewster asserted, once such images are brought "into stereoscopic relief their true character is instantly seen."[20] Further, wrote Brewster, when a stereoscopic photograph was made of important people or events, "the sun will thus become the historiographer of the future, and in the fidelity of his pencil and the accuracy of his chronicle, truth itself will be embalmed and history cease to be fabulous."[21] Despite Brewster's criticism of "spirit rapping," he could not help but attributing to the stereoscope a sublime power to keep the dead among the living.

Brewster was not alone in this celebration of the emotional power of the stereoscope. An American newspaper article from 1858 called attention to "the marvelous fidelity with which [the stereoscope] repeats to us the triumph of man and the wonderworks of God," which were brought before the viewer "in all the sublimity of reality."[22] According to an article in the *Southern Cultivator*, photographs "assume in the Stereoscopic a naturalness and beauty that must be seen to be appreciated, and which possess, in the infinitude of variety, an unfailing charm and delight."[23] For such reasons, claimed a writer in *Godey's Magazine and Lady's Book*, "the stereoscope must rank among the most interesting and most marvelous of modern discoveries."[24]

As was the case with the telegraph and electricity more generally, the same presumed technological power that inspired these celebrations created anxiety for a range of other thinkers. Writing about France in 1859, Baudelaire claimed that with the technology's introduction, "a thousand hungry eyes were bending over the peepholes of the stereoscope, as though they were the attic-windows of the infinite. The love of pornography, which is no less deep-rooted in the natural heart of man than the love of himself, was not to let slip so fine an opportunity of self-satisfaction."[25] An American missionary to China voiced similar concerns that "stereoscopic and other views of the most obscene character are bought from foreigners by peep show men and penetrate hundreds of miles into the interior," and an Indiana statute expressly named stereoscopic pictures as examples of obscene literature. Expressing a different anxiety about the stereoscope's technological power, an English writer worried that

> for our own part, we have left on wondering at anything; nor is there much that strikes us as more marvelous, with our notions of what the spiritual world is, in this than the illusions of the stereoscope, or the likeness we carry in our pocket painted by the light. . . . Some, perhaps, might say that we have touched such an extreme point in the kingdom of nature that we have reached the lowest stair of the kingdom of spirit.[26]

For these critics, stereoscopic technology was both seductive and disenchanting—it gave an overwhelming emotional power to things like pornography, even as it destroyed the enchantment of spirituality.

Although Brewster and these critics disagreed about the social value of the stereoscope, they shared an assumption about its technological and emotional effect. In his discussion of the Stereoscope in *Atlantic Monthly*, Oliver Wendell Holmes argued that when looking through a stereoscope, "there is such a frightful amount of detail, that we have the same sense of infinite complexity which nature gives us." "The stereoscopic figure," he declared, "spares us nothing."[27] Holmes, like Brewster, saw these "frightful details" as beautiful and socially and emotionally enlightening. Still, he was identifying the very characteristics that worried the stereoscope's critics. It was the technologically enhanced details of stereoscopic images that seemed to give them their power to both enlighten and destroy.

It was precisely the presumed emotional power of stereoscopic details that the Keystone View Company and Underwood and Underwood emphasized as they revived stereoscopy at the beginning of the twentieth century. Despite the slump in the stereoscope's popularity in the 1870s and 1880s, by the end of the 1890s these two companies had "cornered the market with millions of views of every country and a multitude of events."[28] Keystone's dominance, in particular, continued well into the twentieth century. It eventually dominated the stereoscope market by "concentrating on the educational sector,"[29] purchasing Underwood and Underwood's stock of stereo-views in 1920. The success of both companies depended on their abilities to link education and enlightenment with the technological and emotional power of stereoscopic photography.

The two companies' sales and marketing approaches tapped into the broader American concerns about technology and emotion that characterized much of the early twentieth century. As I mention in the previous chapter, the associational and technocracy movements of the period offered arguments for how both the government and businesses could better manage the technologies of the period and suggested that emotionally controlled workers would create more efficient industries. The average American experienced these ideas most directly in the form of Taylorism, a method of "scientific management" that found widespread deployment at the turn of the century. Frederick Winslow Taylor was a Philadelphian who revolutionized American factories in the early twentieth century by suggesting ways that they could be better managed. Focusing on the technologies of the new age, Taylor developed elaborate systems for how workers should operate different pieces of machinery, as well as how people and products should move from machine to machine in the course of a production process.[30] The technology theorist Neil Postman saw Taylor's work as "the first clear statement of the idea that society is best served when human beings are placed at the disposal of their techniques and technology, that human beings are, in a sense, worth less than their machinery."[31]

The tenets of Taylorism were not simply confined to the factory. Its advocates "quickly took upon themselves a far greater aim: that of bringing order, rationality, and efficiency out of the disorder, the irrationality, and the wastefulness of the times."[32] In Taylor's view, and in that of his followers, there was a "good that would come to all people if they would live according to the principles of scientific management."[33] Narratives of scientific management made their way into discussions about such varied issues as women's domestic work, American conflicts with Cuba, and the design of libraries and schools.[34] These stories suggested that there was "one best way" to better businesses and better lives, and that this path depended on a level of technological proficiency and emotional self-control. To be a happy, successful, and moral citizen was to lead the "Taylored" life demanded by the new technological age.

Keystone View Company and Underwood and Underwood tapped into these impulses from the turn of the century forward, suggesting that the mastery of stereoscope technology could aid in negotiating the larger tensions of modernity in which many Americans found themselves. Albert Osborne's book *The Stereograph and the Stereoscope*, published in 1909 by Underwood and Underwood, stressed the importance of the optical effects made possible by stereoscopic technology. "It is only in the stereoscope that we get the advantages of two-eye vision," Osborne maintained. "It is true that any picture in which light and shade are properly managed has more or less the effect or *appearance* of space and solidity. But in the stereoscope there is added an entirely different kind of perspective, which, to our eyes, gives perfect depth, perfect solidity, perfect space."[35] Another Underwood and Underwood publication put it similarly: "The prime quality that puts the stereograph in a class by itself is its *depth* or perspective. All other pictures *suggest* depth, but the stereoscope has the far and near of the real landscape."[36] A book published by the Keystone View Company claimed that the stereoscope provides "perfect space for the eye and mind—not merely a suggestion of space as in ordinary pictures; objects stand out in all three dimensions, or as solids, as in nature."[37] Another Underwood and Underwood publication put this still more bluntly. Because of its realism, the author wrote, "the stereograph tells no lies."[38] These claims reiterated a fundamental assumption about the promises of stereoscopic technology. As Jonathan Crary has pointed out, "the desired effect of the stereoscope was not simply likeness, but immediate, apparent *tangibility*."[39]

Both companies offered the tangibility of stereoscopy as proof of its educational value. At the opening of one book, Keystone included testimonies from Charles Eliot, president emeritus of Harvard University, William Bagley, professor of education at Columbia University, and Frank McMurry, professor of elementary education at Columbia University. According to Eliot's published testimony,

> Keystone material provides the means of training children and adolescents to see accurately, to make mental note of what they have seen, and then to put into language whatever has impressed them. It is this combination of visualization with training of the memory, and practice in accurate reproduction of language of what has been pictured in the eye which so strongly commends the method which the Keystone apparatus makes available.[40]

Such testimonies served to reinforce stereoscope companies' claims regarding the abilities of the stereoscopic apparatus to stimulate and educate the mind.

The endorsement of Ivy League professors helped to both legitimize the technological claims being made by these stereoscope companies and to suggest the kind of cultural capital associated with stereoscopy. The fact that college professors endorsed or used stereoscopes would seem to demonstrate the apparatus's status among a certain brand of middle-class intellectual. Indeed, during the early twentieth century, stereoscopes were finding wide use among a range of academics, who repeated the claims about tangibility that were being advanced by commercial stereoscope companies. Arguing for the usefulness of stereoscopic photos in illustrating literary lessons, an English teacher wrote that "the stereoscope shuts the pupil away from outside influences, thus securing a great intensity of impression."[41] For another teacher, the stereoscope was "the most promising help in geography work that is in sight." This was because "with the stereoscope there is impressed on the mind no inconsiderable part of the majesty of grand scenery."[42]

In addition to these classroom activities, geographers and other academics were also using stereoscopes in their own research. In the nineteenth century Francis Galton had suggested that stereoscopy would be especially useful for map making[43] and these claims continued with the twentieth-century repopularization of the stereoscope. For instance, stereoscopic photography became an important form of geographic surveying and reconnaissance for the British military.[44] An author advocating the benefit of the stereoscope to medical research included a series of stereoscopic photos of the various pieces of equipment he described— apparently believing that a "three-dimensional" view would make their technological mechanisms clearer.[45] The stereoscope was similarly utilized in such areas as lunar photography, anatomy, dentistry, and criminal forensics.[46] The stereoscope's ability to make images more "tangible" would apparently benefit any research where contour and depth were important, leaving little outside its "three-dimensional" gaze.

In line with the Taylorist impulses of the early twentieth century, the stereoscope was also used to research and improve people's uses of technology itself. Frank and Lillian Gilbreth, acquaintances and close followers of Frederick Taylor, used stereoscopic photographs in a series of "motion studies" meant to improve the efficiency of various industrial processes. Analyzing workers using a drill press, for instance, the Gilbreths attached a series of lights to the workers' hands and then took motion pictures and timed photographs of them as they worked. A resulting string of light traced each worker's movement and was used by the Gilbreths to determine, and then illustrate, the most efficient series of actions for using the press. These "cyclegraphs," as they called them, were turned into stereoscopic photos—or "stereocyclegraphs"—in order to suggest the multiple dimensions of the movements pictured. These stereocyclegraphs were used by the Gilbreths to create wire models of the ideal set of movements and were also shown to workers themselves. Workers were to imitate the model of the stereocyclegraph, making sure that their own movements traced that of the "motion study man" depicted in their stereoscope.[47]

Finally, the stereoscope was being used and studied by the early twentieth-century psychologists who pioneered much of the era's research on communication technology. Yale's Edward Wheeler

Scripture developed a method of "stereoscopic projection" that, he argued, allowed him to create a simultaneous stereoscopic effect for large groups of people. Using stereoscopic photography and lantern slides—which consisted of a light projected through a glass plate—Scripture created extremely large three-dimensional images. Scripture argued that these projections had a strong psychological power because a three-dimensional image viewed "larger than life size" was "singularly impressive and fascinating."[48] Cornell University's Edward Bradford Titchener likewise stressed the importance of the stereoscope to psychological teaching and research. His psychological laboratory included several stereoscopes—among them an Underwood and Underwood "Perfecscope"—as well as a large series of stereographs.[49]

Not surprisingly, Carl Seashore's laboratory at the University of Iowa—where much early media technology research was taking place—was also analyzing and employing stereoscopes. An Iowa study conducted by James Burt Miner and published in 1905 concerned a young woman who was blind until the age of twenty-two, when a cataract surgery allowed her to see for the first time. Given the emphases at the Iowa lab, Miner was most fascinated by the fact that "Miss W." "had never looked through a stereoscope, opera glass, field glass, or telescope."[50] As he explained, "under these circumstances the experiences of Miss W. with the stereoscope were exceedingly interesting and suggestive," adding that "as soon as it was discovered that she had never looked through a stereoscope, every precaution was taken to leave her completely naïve as to the effect of the instrument." Miner's primary interest was in determining the extent to which Miss W. would be able to experience a stereoscopic effect. The fact that in a short time and with little effort she "readily picked out views with no relief, and with a pseudoscopic effect"[51] seemed to reinforce the natural reality of stereoscopic images.

These psychologists' interest in stereoscopy was not lost on Keystone or Underwood and Underwood. An issue of *Around the World with Burton Holmes*, a magazine published by Keystone, included the following statement under the headline "Nationally Known Psychologists Endorse the Stereograph":

If a stereoscopic photograph of a place is used with certain accessories (as special maps which show one's location, direction, and field of vision, etc.) it is possible for a person to lose all consciousness of his immediate bodily surroundings and to gain, for a short time at least, a distinct state of consciousness or experience of location in the place represented. Taking into account certain obvious limitations, such as lack of color and motion, we can say that the experience a person can get in this way is such as he would get if he were carried unconsciously to the place in question and permitted to look at it. In other words, while this state of consciousness lasts it can be truly said that the person is in the place seen.

The statement was signed by fifteen psychologists, including Edward Titchener.[52] In a statement in another publication, Keystone added another group of psychologists who claimed that when looking at a scene through a stereoscope, "it can be truly said that the person is really seeing the place itself." Among the psychologists presumably endorsing this statement were Yale's George Ladd, George Herbert Mead, and Iowa's Carl Seashore.[53]

In placing an emphasis on the reality effect of the stereoscope and its potential to impact the emotions, Keystone View Company, Underwood and Underwood, and the various academics who endorsed stereoscopy were building an argument regarding its technological sophistication. The stereoscope was largely an outdated, nineteenth-century technology, and the stereoscopic effect was highly variable. As Crary notes, "some stereoscopic images produce little or no three-dimensional effect: for instance, a view across an empty plaza of a building façade, or a view of a distant landscape with few intervening elements."[54] Still, early twentieth-century discussions of stereoscopy managed to emphasize its novelty and technological power. Doing so allowed psychologists, geographers, and other academics to claim access to a high-tech means of representing the various subjects they explored. For Keystone View Company and Underwood and Underwood, it offered a way of framing stereoscopy as an especially sophisticated, enlightened way of looking. Each time viewers experienced the "tangibility" of a stereoscopic photograph in which the reality effect succeeded, they were to feel as if they were looking more deeply into the object or scene it depicted. This added depth, these companies suggested, would in turn stimulate their intellect, emotion, and engagement, all of which would likewise stimulate their sense of citizenship and their individual moral development. According to these manufacturers, the stereoscope was a powerful piece of modern technology offering viewers a transcendent visual and emotional experience.

Three-Dimensional Feelings

The books that Keystone View Company and Underwood and Underwood published to accompany their stereographs were not merely narratives about the images a viewer would see. They were manuals on how to make sense of stereoscopy itself. Among the most important elements of these discussions were instructions on how stereoscopic images should make viewers feel. Here, these companies linked the tangibility of stereoscopy directly to questions of emotional stimulation and development. In *The Stereograph and the Stereoscope*, Albert Osborne called the stereograph "the climax of all in giving the emotions of actual sight."[55] Similarly, an Underwood and Underwood book entitled *Traveling in the Holy Land through the Stereoscope* stressed that the stereoscope presented viewers "not only with life-size representations, but with what are, to a large degree, the actual parts of Palestine itself in their power to teach and *affect* us."[56]

In *China through the Stereoscope*, James Ricalton confronted questions about feeling and experience quite directly in his attempts to convince his readers that stereoscopic travel was just as emotionally enlightening as the real thing. "No traveller brings any material houses or fields back with him," he wrote. "No, the object of the traveller in going so far, at the cost of so much time and trouble, is to get *certain experiences of being in China*. It is not the land, but the experiences he is after."[57] This dualism between mental and physical experience was apparently resolved in the stereoscope, which, through its optical effect, allowed viewers to feel as if they were in China, even when they were home in Ottawa, Kansas.

Osborne's book *The Stereograph and the Stereoscope* offered a sustained argument on the emotional power he identified with stereoscopy. Like Ricalton, Osborne claimed that the distinctions between physical and mental reality were ultimately transcended in the stereoscope. Because of the stereoscope's powerful tangibility, Osborne argued, when looking at a stereograph people experienced a new state of consciousness that was walled off from their immediate surroundings. This made the emotional experience of seeing the images in a travel scene as real for stereoscopic viewers as they would be for actual tourists. As he put it, people "get the emotions of the place or object itself in connection with a picture, to just the degree in which we are able to forget that we are looking at a picture and to think that we are in the presence of the place itself and its surroundings."[58]

This separation of emotion and event was necessary for stereoscope companies to support their argument about the emotional benefits of stereoscopic travel. It was also part of a larger understanding of emotion that had come to prominence at the turn of the century. The psychologist William James, whom Osborne draws on heavily throughout his discussion, had made a famous and seemingly counterintuitive claim about people's emotional lives. Although common sense might assume that people cry because they feel sad, or run because they are afraid, James suggested that the inverse was actually true. "The more rational statement is that we feel sorry because we cry, angry because we strike, afraid because we tremble, and not that we cry, strike, or tremble, because we are sorry, angry, or fearful, as the case may be," he argued.[59] Rather than prioritizing the content of the emotion—say, the intellectual recognition that someone was in danger—what became known as the James-Lange theory of emotion prioritized a person's bodily sensations, placing physiology at the center of emotional experience.

James's emphasis on physiology fit well with the understandings of emotion I identify with media physicalism, and James himself had been heavily influenced by Ernst Mach.[60] In reducing emotion to physiological processes, James imagined emotion in a highly technological way. "Quick as a flash," he wrote, "the reflex currents pass down through their pre-ordained channels, alter[ing] the condition of muscle, skin and viscus."[61] James's description of these physiological processes as "neural *machinery*"[62] further highlighted the technological view that pervaded his approach. In fact, James wrote about the psychology of the stereoscope in his book *The Principles of Psychology*, which even included stereoscopic imagery.[63] Likewise, and despite the fact that James's late nineteenth-century status made him more accepting of introspection and more inclined toward the passions than many of his early twentieth-century successors, his ideas began to suggest the dangers of emotion and the necessity of emotional control. If the emotions were separate from and *preceded* the intellect, then they had the potential to overpower people's rational thought. James's description of emotions as "bodily disturbances"—the term of choice for such later thinkers as Christian Ruckmick—captured well their potentially dangerous consequences.[64]

Osborne likewise presented emotions as something both highly technological and in need of control. While other writers distinguished the mechanical workings of the telephone from the soul melding of the stereoscope, Osborne suggested that the soulful emotional power of stereoscopy was at one with the telephone's electrical transmissions: when a friend "talks into the transmitter of his telephone, . . . certain waves of air come from his lips and strike a thin piece of metal in the transmitter, setting it in motion."[65] As a result of the telephone's technological sophistication, he continued, "these air waves make essentially the same impression on the nerves of our ear as would the air waves from our friend's lips in Pittsburgh." Owing to this, "we can understand how it is, that, in listening to the telephone, our thoughts, our feelings— the whole state of our consciousness is, that we are in the presence not of a machine that gives out articulate sounds, but of a man."[66]

Both telephony and stereoscopy rely on the vibrations of physical waves (sound or light). If these waves carry the emotions of a friend's voice through the wires of a telephone, Osborne reasoned, then why couldn't the technologically sophisticated images of a stereograph carry all the emotional vibrations that one would experience in the presence of the object it photographed? Throughout this and other sections of his book, Osborne describes in great detail both the technical apparatus of the stereoscope and the physiology of binocular vision, trying to draw a scientific picture of the paths by which these vibrations pass. If the stereoscope had a powerful impact on people's emotions, Osborne claimed, it was because of the essentially technical and physiological status of the emotional stimulations it made possible.

Like James, Osborne also suggested that emotions needed to be controlled in various ways. In terms of the stereoscope itself, Osborne argued that one needed to use it "systematically" and in highly formalized ways in order to receive its benefits. This meant following along carefully with the included books, being sure to orient oneself with the included maps, and in general avoiding the "bodily indolence" that typified most photographic viewing. Drawing from James's ideas, Osborne wrote that "it follows that if we want to experience a certain state of feeling, we can do much to bring it about by assuming in advance the appropriate bodily attitude."[67] Viewers needed to exercise a certain manner of bodily and emotional control in order to receive the emotionally enhancing benefits of stereoscopy.

Finally, and ironically, Osborne suggested stereoscopic travel as a kind of antidote to the emotional and technological chaos of the early twentieth century. Reflecting the basic ideas about information overload that would be voiced by Wallas, Lippmann, and others, Osborne stressed the dangers of this sort of chaos to people's emotional well-being. "We must strive to get in touch with and keep in touch with the environment that will tend to give us the desirable thoughts and feelings, and to keep away from the environment that would give us undesirable thoughts and feelings," he wrote.[68] In the contemporary world, Osborne claimed,

> even most of the millions gathered in cities live truly narrow lives. The vast majority of their days they tread the limited round from the home to the office or shop, and from the office or shop to the home. In rural life the great drawback is the meagerness and narrow range of one's experience and impressions; in cities the danger is that the person's attention will be so taken up by the multitude of commonplace impressions that he has little time for the more worthy objects of attention.[69]

In taking control of their own emotional lives via the stereoscope, users would presumably gain a high-tech solution to the emotional chaos of the information age.

Feeling White and Middle-Class

As Osborne's ideas suggest, in the new technological age, mastering technology became an important component of self-mastery more generally. A "Great Society" overloaded with information and whizzing automobiles demanded citizens who could resist communicative and technological overstimulation even as they took advantage of the modern possibilities these developments offered. In connecting stereoscopy to questions of emotional development, Keystone View Company and

Underwood and Underwood worked to frame the supposed emotional stimulation of the stereoscope as an especially positive one. It was a technological path toward a more emotionally controlled self and, with it, an enhanced social standing.

These arguments about the stereoscope fit into a larger climate of early twentieth-century consumption inflected by anxieties over white middle-class identity. The turn of the century saw a rapid increase in geographic and social mobility, against which the virtual mobility of the stereoscope made technological and cultural sense. From the late nineteenth to the early twentieth century, a rising number of immigrants tested the boundaries of the nation's democratic promise. From 1870 to 1920, the population of the United States born in foreign countries rose from 5,567,234 to 13,920,692—a 250 percent increase.[70] Racist tensions mounted during this period. The Chinese Exclusion Act of 1882 sought to limit immigration, as did subsequent acts targeting Asia more generally as well as those of African dissent. These decisions culminated in the National Origin Act of 1924, which aimed to restrict immigration to white western and northern Europeans.[71] Even still, the status of Italian, Irish, and Jewish immigrants and others of "variegated whiteness" was not guaranteed. "White" was a category reserved for people of a particular bodily and emotional comportment.[72]

Alongside these racial tensions, the changing economic conditions of the country created a set of dynamic class relations as well. Improvements in manufacturing made mass-produced clothing and other personal and household items available to a new class of consumers distinct from the laboring classes who manufactured them. This industrial growth also gave rise to new managerial positions designed to mediate between wealthy industrialists and their working-class employees.[73] By the end of the nineteenth century, "the manual-nonmanual basis of work clearly differentiated workers with respect to their opportunity for stable membership in the middle class."[74] For Frederick Taylor, there was a distinction between workers who used industrial machines and those who supervised them using the more intellectual tool of the slide rule. For instance, the "science of running a lathe," Taylor claimed, was "so intricate that it [was] impossible for any machinist who is suited to running a lathe year in and year out either to understand it or to work according to its laws without the help of men" who were trained in the slide rule's more complex mechanics.[75] Taylorism identified the new middle class with a set of technologies that were more removed from the hands-on equipment of the shop room floor.

In linking stereoscopy to a kind of cultural advancement, stereoscope companies tied their technologized vision of emotion to a particular kind of class and racial standing. As I mention above, photography had acquired an important place in constructions of race as part of eugenics, Francis Galton believing that the careful application of photographic technology could provide a scientific image of the basic characteristics of different races and classes. Eugenics assumed that people's personalities, temperaments, and so forth were hardwired into their bodies. In fact, Galton employed stereoscopic photos in some of his composite portraiture, suggesting that it gave him a three-dimensional vision of the traits of a given character type. From the standpoint of eugenics, changing the presumed problems in a given group of people required the kind of genetic engineering that could be accomplished only across generations.

Despite the physiological focus of their ideas about emotion and their belief in higher civilizations, these stereoscope companies did not advocate a eugenics position. Indeed, for stereoscopic education to work, people's *environments* needed to play a fundamental role in their emotional makeup. Osborne took great pains to argue that people's characters are not determined by their inner makeup, but by the kinds of stimulations that are available to them. "Man does not lift himself by his boot straps in any sense, physically, mentally or morally," he wrote. Rather, "he must have something upon which to climb, . . . [and] the height at which he attains will depend largely on that which serves as his ladder."[76] Of course, placing an emphasis on environment did not free these companies from their arguments about the advancement of civilization, as Osborne further indicated:

> We can easily understand how it is that an infant placed in some secluded part of China today and brought up apart from all western influences must grow up to be a Chinaman in all essential respects in thought and conduct. Certainly it would be utterly irrational to expect him to develop into a representative of modern civilization and enlightened Christianity, by the exercise of any power he possesses within himself.[77]

The quality of a civilization depended not on its people's genetic makeup, but on the quality of their ladders—or whatever technology would presumably elevate them culturally and emotionally.

In stressing environmental factors over genetic ones, Keystone View Company and Underwood and Underwood more clearly reflected a different—but no less creepy—early twentieth-century movement: *euthenics*. In her book *Euthenics, the Science of the Controllable Environment*, Ellen Richards defined the term as "the betterment of living conditions, through conscious endeavor, for the purpose of securing efficient human beings."[78] While eugenics focused on "race improvement through heredity," euthenics emphasized "race improvement through environment." As such, wrote Richards, "euthenics precedes eugenics, developing better men now, and thus inevitably creating a better race of men in the future."[79] Among the early twentieth-century figures championing euthenics was Carl Seashore, who promoted the concept as early as 1927 and continued to champion it throughout much of his career.[80] To those committed to the idea that appropriate uses of technology could enhance people's mental and emotional life, Richards's claims that "scientific knowledge" could improve both individuals and the whole of a civilization no doubt made perfect sense. It also, of course, justified a highly stratified conception of civilization that saw less advanced cultures as technologically and emotionally—if not necessarily genetically—backwards.

In keeping with this basic rhetoric, as Keystone View Company and Underwood and Underwood connected the emotional power of stereoscopy to various sorts of travel experiences, they placed a range of limits on just what constituted appropriate emotional stimulation and technological use. If the emotional experiences of stereoscopic travel were to elevate people above the emotional cacophony of the early twentieth century, they needed to offer up a particularly refined emotion. If stereoscopic technology was to offer an escape from the technological saturation of the age, it would need to be an especially refined technology. Stereoscope companies were promising a particular media physicalist citizenship that offered a narrowly white and middle-class manner by which people could achieve technological and emotional self-control.

As might be expected, Washington, DC, was a popular theme for stereoscopic travel in that it suggested the kind of personal betterment that could come from one's sentimental connections to the nation.[81]

The 1904 book *Washington through the Stereoscope* focuses explicitly on the *emotional experience* of viewing the city stereoscopically. Discussing the book and its accompanying set of stereographs, the testimonial's writer holds that "a more stimulating and quickening aid to education cannot be placed in the hands of the people, especially of our youth."[82] Washington is likewise "the heart and nerve-centre of our national life, endeared by a thousand moving and heroic associations to uncounted millions of men."[83] Explicitly instructing viewers on the sublime experience they should expect from the Washington Monument, the author calls one image "truly a fitting setting for the noble column which each moment claims a larger mede of our awe and admiration, for the Monument is like a mountain in that it grows on its beholder. . . . You will admit that I spoke truly when I said that it was the most imposing single object of great dimensions erected by modern hands."[84]

In the artistic beauty of Washington—experienced three-dimensionally—these stereoscopic tours suggested that spectators could have direct emotional experience with the nation's heroic past. In narrating another stereo-view, the tour guide for *Washington through the Stereoscope* comments, "It was across Long Bridge down yonder that Julia Ward Howe drove on an autumn day in 1861 for the visit to a review of the Army of the Potomac, encamped on the Virginia Hills, which gave birth to her match-less Battle Hymn of the Republic." The next page contains the poem's full lyrics, which, according to the author, "have in them the very breath of a heroic time, and of the feeling with which it was filled."[85] Arlington Cemetery made a popular subject for these stereoscopic photographs as well. One Keystone stereoview, dated December 28, 1899, captures the burial of the victims of the battleship *Maine*. The picture depicts rows of flag-covered coffins, diagonal to the viewer, stretching out into the far background of the picture. The picture's photographic perspective, in which the coffins shrink in size the further they are from the camera, takes advantage of the stereoscope's three-dimensional effect, suggesting that the fallen bodies go on ad infinitum.[86] "On the level plateau in front of us," reads the caption for another stereo-view of Arlington in *Washington through the Stereoscope*, "the headstones of white marble stretch away in lines seemingly endless to the vision."[87] Presenting these images in their three-dimensional glory, these stereoscope companies promised a sublime access to the concrete history of the nation's fallen—something that many stereoscope consumers would not have been able to experience "firsthand."

In addition to instructing viewers on how to experience the emotions of spectatorship, these companies also suggested how they should make sense of technology itself. One Keystone stereograph dated 1917 depicts four high school students of McKinley Manual Training High School in Washington, DC. The photograph, entitled *Doing Their Bit*, shows the four male students holding munitions shells they have just made.[88] Another stereo-view from the same year depicts a group of young women at a National Service Camp for Girls in Washington being educated on the use of the radiophone.[89] Yet another image depicts one hundred different workers at Washington's Bureau of Printing and Engraving setting type on monotype machines.[90] Not unlike commentaries on the stereoscope more broadly, these images draw explicit links between technology and citizenship, suggesting that a person's mastery of technology plays an important role in his or her personal development and service to the nation.

If these images of laborers depict one version of modernity, a pair of Underwood and Underwood images titled *Pennsylvania Avenue from the Treasury, N.E. to the U.S. Capitol,*[91] illustrates yet another, linking stereoscopic travel with more cultivated, civilized sensibilities. Both photographs depict a small group of people standing against a carved stone railing, looking down on and across Pennsylvania Avenue toward the Capitol Building. The women wear blouses, long skirts, and wide-brimmed feathered hats. The men wear dark suits and hats, one sporting a hard straw, flat-topped "skimmer" or "boater" hat and the other wearing a dark "bowler" or "derby." In depicting this particular group of people, Underwood and Underwood clearly represented the leisure class that was developing around the turn of the century. The man in the skimmer hat stands sideways, with his left arm leaning on the railing in front of him, his left knee casually bent. Similarly, the man in the bowler stands with both of his elbows on the railing, apparently relaxing as he looks down the avenue. Two women lean on the railing similarly to the men, casually propping themselves up with their elbows. Another woman sits on the railing with her legs dangling in front of her. Both their relaxed postures and their clothing, which contemporary viewers would have recognized as upper-middle-class

Figure 2.1

Doing Their Bit: Students of Mckinley Manual Training High School, Washington, D.C., with 4-Inch Shells They Have Made (Meadville, PA: Keystone View Company, 1917). Courtesy of the Library of Congress.

Figure 2.2

Class in Wireless: National Service Camp for Girls, Washington, D.C. Outdoor Class (Meadville, PA: Keystone View Company, 1917). Courtesy of the Library of Congress.

leisurewear,[92] suggest that these people have the time and money to spend leisurely looking at this Washington monument.

With the leisurely spectators set in the extreme foreground and the Capitol Building at a distance away, this image takes clear advantage of the stereoscopic perspective, depicting this group of onlookers as "larger" and thus apparently "nearer" the viewer. As the image took three-dimensional form through their stereoscope, contemporary viewers would have found themselves on the balcony with these fellow tourists, apparently gazing with them at the distant Capitol Building. Through this and similar perspectives, Underwood and Underwood encouraged viewers to identify with wealthier, presumably more culturally developed tourists. Through their stereoscope, these companies suggested, viewers could experience Washington in the same manner as the more leisurely members of the higher classes. In addition, these

Figure 2.3

Pennsylvania Avenue from the Treasury, N.E. to the United States Capitol (Washington, DC: Underwood and Underwood, 1903), graphic. Courtesy of the Library of Congress.

companies' repeated discussions of stereoscopic technology framed stereoscopy as a uniquely modern viewing practice, itself appropriate for a modern, mobile, middling class.[93] As they came into three-dimensional view, these leisure-class tourists and the Capitol on which they gazed served to remind the viewer of the uniqueness and modernity of the technology before them. As a consumer of a stereoscope, the at-home viewer was a technocrat with access to the cultural experiences and cultivated emotions of more moneyed tourists.

If domestic stereoscopic images provided views of high-tech, enlightened citizens with whom viewers were supposed to identify, international views provided images of various others against whom this new emotional and technological citizenship could also be developed. On the one hand, these international scenes were educational resources through which the stereoscopic traveler could expand his or her

emotional development, the descriptions of foreign sites often rivaling the above discussions of Washington. In discussing the view "from the dome of St. Peter's" in Rome, one writer asserted that "all must admit that this colonnade enfolding the Piazza is imposing, almost sublime. There is nothing equal to it in any temple in the world."[94] Evoking a similar sense of sublimity, another writer focused on the size and aesthetic power of the Great Pyramids: "If you will let your eyes run down the precipitous sides nearly 500 feet to the desert below you will hardly be ready to shrink back, I doubt not, at the suggestion of falling."[95]

Even as such descriptions identified an emotional power with these foreign monuments, others highlighted the backwards emotional and technological status of the people living alongside them, creating foils against which the stereoscope viewers' more supposedly cultured status could stand out. Focusing on the emotions of the people in these other countries, the tour guide in *India through the Stereoscope* told his readers that the Indians they saw through their stereoscopes were engaged in "a universal struggle for a miserable existence."[96] The same writer who praised the colonnade enfolding the Piazza said of a stereograph entitled *The "Lazzaroni," as They Live in the Streets of Naples*, "What a scene for degenerate character study!" According to the narrator, these people were originally a "semi-criminal class" that contained "many vicious criminals."[97] Likewise, he described an allied group, the "Camorra," as "a class of ruffians *addicted* to all degrees and variety of crime."[98]

In discussing a photograph of a mission school in Shanghai, James Ricalton appeared to compliment the emotional demeanors of a group of Chinese students, only to turn it into a celebration of Western emotional development:

> I desire especially to call your attention to their bright faces. On several occasions before I have asked you to notice the sad and expressionless faces in native groups; but these countenances are scarcely more than half Mongolian; they are bright and cheerful. . . . All are neat and tidy, and from refined homes and under a faithful American teacher.

Ricalton offered a blatantly physicalist explanation for their bright faces: "This is partially owing to the fact that they have just been looking through the stereoscope."[99] Whereas stereoscope viewers felt the educational powers of the sublime, the foreign people on whom they gazed were sad, expressionless, miserable, vicious, and addicted.

In a similar way, these stereoscopic images and commentaries highlighted the supposed inferiority of the technologies used by the people in these lands. One of Keystone's images, entitled *A Block of Tenements in Which Some of China's Floating Population Dwell, Hong Kong, China*, highlighted the apparently low-tech nature of Chinese living conditions, a long line of wooden boats, paddles, poles, and ropes, jumbled together in an apparently chaotic scene.[100] These ideas were further reinforced in

Figure 2.4

A Block of Tenements in Which Some of China's Floating Population Dwell, Hong Kong, China (Meadville, PA: Keystone View Company, 1906), 1 photographic print on stereo card. Courtesy of the Library of Congress.

travel commentaries that repeatedly emphasized the country's supposedly low-tech, nonscientific ways. A discussion in one Keystone travel
publication explained that "Chinese boatmen believe in water demons
and if a boat comes to grief by collision, nothing will be done by their
neighbors to save the life of the crew," because "any attempted rescue
might bring the wrath of the demons on themselves." Likewise, because
the boats were believed to be guided by spirits, the article reported, many
were adorned with "hideously carved" eyes.[101] Such descriptions posed a
strong contrast with the high-tech, "scientific magic" of the stereoscope.

Other images and commentaries similarly focused the viewers' attention on the technologies of these foreign lands. The narration for an
image entitled *Eskimo Girls in the Frigid Arctic, Cape York, Greenland,*
noted the clothing of the women it depicted: "The skins are scraped and
worn until they are very soft and pliable, and are sewed with a thread of
sinew and a bone needle."[102] Discussing an image of a mother and her
child outside a hut in New Guinea, another commentator explained that
"the queer white thing sticking through her nose is bone. The girl has one
too. They wear them to make themselves look handsome! Perhaps you
think they succeed."[103] A stereograph of Seoul, Korea, explained how "the
little, low houses are thrown together regardless of streets," adding that
"the furniture is simple and made of bamboo like that of the Japanese."[104]

These disturbing discussions of race followed directly from stereoscope companies' assumptions about emotion and technology. New
technologies—or even supposedly new technologies, such as the stereoscope (Osborne devoted a good deal of space to why the stereoscope
disappeared in the late nineteenth century, arguing that it had been too
sophisticated to be understood in that earlier time!)—had a particular
power over the emotions. Emotions were transmittable, physiological
processes that could be both culturally enhancing and detrimental to
intellectual development. Precisely how people developed depended on
the technological and cultural conditions of their environment. Thus
people who used such an emotionally and culturally enhancing technology as the stereoscope would be changing themselves for the better,
while those without this, or even seemingly less emotionally enhancing
technologies, would be at the lowest rungs of the ladder of civilization.

All these ideas were supposed to come together when a person
took a stereoscopic tour, experiencing a collection of stereographs,

accompanying narratives, and manuals of technological and emotional instruction. As images of monuments, leisure-class tourists, floating tenements, and "Lazzaroni" snapped into three-dimensional view, this stereoscopic rhetoric promised, the cultural capital of stereoscope viewers would be enhanced in at least two ways. On the one hand, viewers were to feel a kind of techno-emotional enhancement in which their bodies, their emotions, and their intellect achieved the kind of transcendent harmony made possible by the tangible images of the stereoscope. On the other hand, and in contradistinction to the low-tech foreigners they gazed upon, these tourists were to feel the enhanced standing that came with the ownership of the high-tech modern media apparatus of the stereoscope itself.

Emotional and Technological Convergence

The affective scope of stereoscopic tours was widened by the multiple discourses they drew together. Some of these, such as the books published to accompany stereoscopic pictures, were under the companies' control. Others, such as the academic commentary that celebrated the stereoscope—especially that which took place in the context of specific geographic, psychological, or other research—and the larger cultural conversation about the apparatus, took place outside these companies' domain. Together, these various discussions created a larger rhetorical ecology in which to make sense of the stereoscope's presumed technological and emotional power. Stereoscope companies were tapping into—and perpetuating—a set of ideas about the technological force of the stereoscopic effect, as well as larger anxieties about modernity.

The technological and sentimental rhetorics being built by these companies included still more than their books and stereoscopic photographs, however. They were building larger *social networks* that included the discourses of stereoscope salespeople and consumers themselves. Combining their virtually real views of the world, travel commentaries, face-to-face interactions with stereoscope salespeople, and personal travel newsletters, Keystone View Company and Underwood and Underwood worked to construct a fully immersive rhetoric through which they could sell their wares. Consumers were expected to feel a real connection to the emotional scenes depicted in their stereoscope as well as to stereoscope companies themselves.

The travelling salesman—the gendered term of choice for both Keystone View Company and Underwood and Underwood—was an important node in this stereoscopic discourse network. Predominantly college students, teachers, and similarly middle-class, educated folks, stereoscope salespeople were given a specific territory—usually a geographic region surrounding their college or hometown—over which they were authorized to sell a whole gamut of stereoscopic books, views, apparatuses, and other paraphernalia. Keystone told one salesman that if he committed himself to his work, he could expect commissions from fifty to seventy-five dollars per week.[105] However, both Keystone View Company and Underwood and Underwood maintained, successful salesmen would need to be more than just stereoscope advocates. They were expected to embody the same emotional, moral, and social well-being that was presumably developed through stereoscopic viewership.

Salesmen were expected to comport themselves in a way commensurate with the claim that the stereoscope was both educational and entertaining. In one sales manual, Underwood and Underwood instructed its salesmen to "rise to the full importance and dignity of the work," and Keystone insisted that a certain "earnestness" was central to successful stereoscopic salesmanship.[106] Underwood and Underwood's manual offered three different sales talks and a collection of responses to various interruptions and objections that potential buyers might make, all of which were to be memorized. These salesmen also needed to learn facts about the various places depicted in the tours they were selling, so they could model the "travel background" these sets promised. While relaying this cultivated knowledge, salesmen were also expected to be "enthusiastic" and "interesting," to "keep in good spirits," and to be "thoroughly alive to the work."[107] Both cultured and fun, the salesman was to personify the sort of educated, leisure-class subject depicted in these companies' stereoscopic tours of Washington, DC.

In order to be successful, these salesmen needed not only to cultivate and control their own emotions, but also to analyze and educate the emotions of their customers. A salesman was to learn to explain the arguments about tangibility and emotional development that these companies developed in their books and other literature (Underwood and Underwood's manual, in fact, advised salesmen to read Osborne's book *The Stereograph and the Stereoscope*). Because they made face-to-face contact with customers, salesmen were in a unique position to educate customers about stereoscopy. Following Underwood and Underwood's sales talks, salesmen would lead viewers through a series of stereographic tours, drawing their attention to particular elements of images that highlighted the stereoscopic effect. For instance, salesmen were instructed to show customers an image outside the stereoscope before having them look at it through the viewer, at which point they would call attention to details most enhanced by the three-dimensional view. Ultimately, the salesmen would go on to explain, travelling was less about external realities than it was about "internal, non-material thoughts and feelings," and it was just these "inner experiences" that were stimulated by the stereoscope.[108] Salesmen were expected to teach their customers to *see* and *feel* the power of stereoscopic images.

If salesmen offered one kind of contact between stereoscope companies and their customers, Keystone's "travel club" provided yet another. Conceived in the 1920s, the travel club was primarily focused on school-age children, but was targeted to older stereoscope customers as well. Members of the club received a monthly "travel magazine" and a series of stereographs depicting various scenes described in the magazine. One of the magazine's covers depicted a boy and girl looking at stereoscope pictures before a mosaic of global imagery, all above the caption, "We See, We Learn, We Experience." The club and magazine created an interactive means through which Keystone could promote the powers of stereoscopy. In support of the stereoscope's presumed educational benefits, Keystone provided rewards to students who improved their grades after "studying" their stereoscopic tours. Students would mail in a "certificate of improvement" signed by their teacher and would then receive watches, stereographs, and other prizes.

The "face" of the travel club was "Anne Travelog," the club's supposed secretary, who encouraged readers to write in about their various experiences with the stereoscope. Issues of the travel magazine regularly featured "personal letters" from Ms. Travelog, in which she addressed the club members directly, encouraging them to remain active and engaged. As she wrote in one letter, "remember this is YOUR CLUB, YOUR MAGAZINE, and every member may have a part in making it a success."[109] Each issue of the travel magazine also included a set of questions from Ms. Travelog that readers were supposed to fill out and return.

Figure 2.5

A cover image from the *Keystone Travel Club* magazine. Courtesy of the Johnson-Shaw Stereoscopic Museum, Meadville, PA.

One issue included such questions as "How long did it take Lindbergh to fly across the ocean?"; "What two rivers flow together to form the Ohio River?"; and "What is Pig Iron?," the answers to which could be found in a series of Keystone stereographs.[110]

Written responses from club members were another important feature of both the *Keystone Travel Club* magazine and their other travel club monthly, *Around the World with Burton Holmes, America's Premier Traveler*. These responses reinforced Keystone's claims about the importance of club members and gave those members an opportunity to support the arguments offered in the company's writing about the stereoscope. "This is the most marvelous photography I have ever seen," wrote T. L. Black, president of the ACME Coffee Company of Fort Worth, Texas. "I cannot express my gratitude to the man who came my way and convinced me that he had something DIFFERENT, which really is true."[111] The same issue included a letter from Hugh Forman, who reported that Douglas Fairbanks had shown Forman his Keystone Library. Fairbanks had apparently explained that the stereographs were "consulted as the final authority before the 'sets' for any foreign picture are made." Forman concluded that "the 'educational' value of the moving picture evidently depends upon the stereograph."[112]

These magazines also frequently included letters and other writings from student club members. Some, such as eight-year-old Betty Lou Hall, wrote to explain how much their stereoscope had helped them in their schoolwork.[113] One issue included a poem by twelve-year-old Stephen Arnold Goldstein celebrating the magic of stereoscopic travel ("No carpet that sails through the air, No lamp of Aladdin, or Seven League Boots, But a Magic that needs no great care"). The poem's last stanza captured the basic rhetoric that Keystone had worked hard to promote:

> If the day is so hot you can't stand it at all,
> Choose the Alps where the snow's to your knees—
> You'll stand on the top of an ice-covered peak,
> And shiver and chatter, and freeze.
> If the weather outside is dreary and cold,
> You need never be gloomy, I'm sure;
> You can transport yourself to a sunnier clime,
> In a telebinocular tour.[114]

Another issue included an essay by Lillian Bell, of High Rolls, New Mexico, which had won second prize in Keystone's fifth-, sixth-, and seventh-grade essay contest. In the essay, the girl's fairy godmother led her on a tour of the world, via her Keystone stereoscope.[115] As with David Brewster, the sense of the natural magic of stereoscopy was strong with these young viewers.

If the argument about the stereoscope's tangibility was getting through to club members, so was its status as a white, modern, middle-class technology. Nine-year-old June Tracy of Milwaukee, Wisconsin, sent Keystone a letter that included a fourth-grade report she wrote on "Eskimos," which was subsequently published in the *Keystone Travel Club* magazine. Echoing the narration included in Keystone's *Eskimo Girls in the Frigid Arctic*, Tracy highlighted what she saw as peculiarities about Inuit life, focusing in particular on their low-tech lifestyles. "Only rich little boys have wooden sleds," she reported, "otherwise they have sleds of snow and ice." "They have no clock," she continued, adding that "perhaps if they ever saw one they would be afraid of it." Stressing their presumably superstitious nature, she further explained that "the Eskimos believe in giants and magic. They believe that there is a big giant right in the middle of their land."[116] Of course, such superstitions were no match for the high-tech, scientific magic of the stereoscope through which Tracy learned her geography.

Tangible Feelings

Stereoscope companies benefitted from and capitalized on a whole host of ideas about technology at play in the early twentieth-century United States. In addition to specific arguments about stereoscopy made in academic and popular sources, the Taylorist movement, concerns about emotional control, the growth of the middle class, and anxieties about immigration all seemed to uphold the importance of a certain technological and emotional ethic. As framed by Keystone View Company and Underwood and Underwood, the three-dimensional effect of the stereoscope was the perfect apparatus for this modern period of middle-class self-control. Indeed, because most people do not immediately see an image in three dimensions when looking through a stereoscope (despite claims about the device's natural perspective), users needed to *learn* how to take advantage of this high-tech device. Once their eyes adjusted, and the stereoscopic effect took form, they would have made an ocular correction that—the rhetoric surrounding the device maintained—opened them up to a newer, more sophisticated form of vision. Stereoscopic subjects personified modernity intellectually, emotionally, and in their very bodies.

The comments of club members (the Keystone Travel Club remained active until 1958), and the successful sale of stereoscopes more broadly, suggest that this stereoscopic rhetoric was having an influence on the culture at large. However, the impact of this network of discourses was more important than simply getting people to purchase stereoscopes. No doubt many stereoscope owners never joined a travel club or imagined themselves to be travelling by magic carpet when they looked at a stereograph. Still larger numbers never bought a stereoscope in the first place. But in its cultural and academic success and the publicity it garnered as a result, stereoscopy advanced the larger rhetoric of media physicalism that was taking hold at the beginning of the twentieth century. The link between the "tangibility" of a representation and the emotional being of an audience would pervade discussions of the new mass media as a whole.

Likewise, assumptions about the "educational" quality of the early twentieth-century stereoscope had important parallels in the other mass media of the time. As some of the testimonials above demonstrate, both teachers and stereoscope companies emphasized how stereoscopy could be useful in the classroom. Created in 1905, Keystone's "600 Set" was a collection of stereographs targeted specifically to schools. An accompanying book, *Visual Education through Stereographs and Lantern Slides*, explained how the images could be used to teach everything from geography and history to government, spelling, and mythology.[117] Those schools that did incorporate stereoscopes provided an important market for Keystone and other stereoscope companies. Not only did they buy stereoscopes and stereographs themselves, but they publicized stereoscopy among a group of students who might themselves become stereoscope consumers. Just as importantly, the integration of stereoscopes in schools helped establish a trend of educational uses of the new mass media; phonograph records, radio programs, and motion pictures were also marketed to, and found their way into, schools throughout the country. Many educators became convinced that a good education required the use of the newest communication technologies—reflecting the sorts of arguments about technological "tangibility" and intellectual stimulation that stereoscope companies were pushing so heavily.

Finally, the ideas about class and race that were important to the success of the stereoscope would find expression around phonographs, radio, film, and the other new media as well. Anxieties about modernity were largely concerns about how individuals could adjust themselves to a new, more complicated, more diverse, high-tech world. While new technologies were exacerbating these anxieties—for instance, by making pornographic images available, or by threatening people with emotional overstimulation—they also promised sophisticated ways to transcend their corrupting influences. Stereoscopic travel offered viewers a virtual escape from their location and their selves, creating an elevated view from which they could look down on the more presumably low-tech people from whom they were distinguished. To be middle-class, modern, and American was to be a consumer of the stereoscopic apparatus and of the ideal, emotionally and technologically sophisticated citizenship attested to by its gaze.

3

Electrifying Voices

Recording, Radio, and the New Friendly but Formal Speech

Throughout the eighteenth, nineteenth and early twentieth centuries, practitioners of elocution had sought to both entertain and educate listeners regarding the emotional possibilities of the human voice. Virgil Pinkley's 1897 book *Essentials of Elocution and Oratory* promised readers "vocal and physical equipment for the purpose of speech, the greatest gift of God to man." As Pinkley explained,

> When breath, body, and voice are made subservient to the mind; when the mind is made to know what are the demands of thought; when the emotions are in keeping with the character of the thought; when all these forces act in harmony with the requirements of the thought, then has the Art of Elocution and of Oratory touched its zenith.[1]

Pinkley went on to spell out the essential components of elocutionary practice, including voice preparation, gesture, calisthenics, and emotional expression. Elocutionary study offered students a means of disciplining their minds and bodies and presenting a more cultivated overall personality.

Given this emphasis on oral expression and performance, it should hardly seem surprising that two of North America's most prescient thinkers about communication media were both the children of elocutionists. Alexander Graham Bell's grandfather was a gifted orator, and the telephone inventor's father, Alexander Melville Bell, was a pioneering elocutionist who developed a phonetic alphabet to aid the deaf. Marshall

McLuhan's mother, Elsie, found wide success as she travelled across Canada for elocutionary and oratorical performances. She also taught elocution to a number of students, including her son Marshall, whose own affinity for oral culture, his frequent quotations of poetry, and his bombastic voice bore the imprint of these lessons. The elocutionists offered their own early versions of media theory. They aimed to perfect the human body as a medium for the expression of emotion via the voice.

As Marshall McLuhan might have predicted, developing recording technologies had a profound influence on the speech practices in which the elocutionists engaged. Like visual technologies such as the stereoscope, the recording apparatuses of the early twentieth century impacted not only everyday consumers, who could now own and replay their favorite songs and other sounds, but a range of scientists, scholars, and businesspeople who tried to make sense of these new technologies and adapt them to their work. The ability to record sound allowed scientists to analyze material that had simply disappeared in the past, and thus opened up a number of new research areas. Among the most important of these was human speech itself, the speech scholars of this period employing a range of the era's new technologies in their research. By using recording technologies, early twentieth-century scientists believed they could develop a much deeper understanding of the human voice. The early twentieth-century United States saw important changes in research and teaching about speech, as a variety of voice experts tried to adjust these vocal practices to account for the new technologies through which they captured them.

Into the early twentieth century, American speakers had continued to practice a wide range of styles, with the highly emotional style of elocution figuring very prominently among them (in fact, in 1888 when Edison enumerated the uses to which his new phonograph could be put, "the teaching of elocution" ranked third after letter writing and phonographic books; the reproduction of music ranked fourth).[2] However, by the 1920s, speech scholars as well as others engaged in the instruction and practice of speaking would advocate a more presumably emotionally controlled, natural model of speaking. This transition was indicative of the series of tensions regarding communication, technology, and emotion that arose during this time. As the previous chapters illustrate, the presumably sublime powers of the new technologies of the twentieth century caused both great joy and great concern regarding the emotions. The supposed high-tech power of the stereoscope could deliver the emotional experience of visiting the Holy Land while the spectator sat at home on his or her couch. This made it ideal for scientific and educational practices, but also threatened the emotional overstimulation that worried a number of important early twentieth-century figures. The new public speaking of the 1920s reflected these conflicting hopes and fears regarding the ability of recording technologies to capture and transmit human emotions.

This chapter explores the rise of this new model of speech, with a particular attention to the role of speech teachers and other assumed authorities in its development and dissemination. The first section places the new public speaking in the context of the emotional climate of the 1920s. The speech discipline's rejection of elocutionary practice was part of the larger trends toward emotional control that pervaded many aspects of the period. The second section considers why teachers and scholars may have found solace in new, technology-centered ways of thinking about speech. For speech, as with many other disciplines in the 1920s, the shift toward technology promised a more rationally focused research that fit with the period's broader culture of emotional control. The third section considers one of the most powerful ways that these debates about speech, technology, and emotion entered the public: through discussions surrounding the radio announcer. Radio speech highlighted some of the period's hopes and anxieties about the new technology's ability to reproduce and amplify the human voice, giving public shape to the issues explored in the speech laboratory. The final section demonstrates some of the tensions about race, class, and gender that, as they did in discussions of the stereoscope, attached to these debates about technological and emotional power.

Together, speech scholars, journalists, media producers, radio announcers, and other assumed experts on the voice built upon and further developed the era's broader rhetorical ecology of media physicalism. In measuring the emotions of speech through various technologies, speech scholars presumed that they were stripping away unnecessary ornamentation in order to arrive at a more natural understanding of the voice. This allowed these researchers to place emotion at arm's length. Emotions were technical issues to solve scientifically, rather than the more impassioned and public displays that characterized elocution. By the same token, the science of speech could be presented as both a highly practical exploration of the everyday voice and a highly sophisticated, technologically advanced kind of research. This same conception of *everyday sophistication* was at the core of the new public speaking as well. Faced with the cacophony of chaotic voices that were believed to characterize the Great Society, speakers were to restrain their own emotions—to sound "natural"—even as they sought technologically sophisticated ways to amplify their voices above the crowd. In this way, the new public speaking held much in common with the rhetoric that sustained popular stereoscopy, although rather than necessarily owning a particular technology, the new speaker was to perform a kind of technological aesthetic by emulating certain features of the new recording technologies in his or her own vocal and bodily practices. Likewise, concerns about "naturalness," "friendliness," and "personality" illustrated the same set of largely white, middle-class, and masculine norms about emotional and technological control that dominated stereoscopy. Like the stereoscope viewer, the public speaker, radio announcer, and speech scholar were supposed to perform very similar kinds of technological and emotional citizenship, embodying a kind of narrow, emotionally controlled, technology-centered personality that seemed demanded by the new media era.

The New Emotion and the New Speech

As I discuss in chapter 1, Peter Stearns identifies the emergence of the general cultural climate of "American cool" with the period of the 1920s. Western culture had long displayed a broad apprehensiveness about emotional expressions—going back at least as far as Socrates—and the period of the "civilizing process" identified by Norbert Elias had seen a further emphasis on emotional control, especially for the middling classes. The period of the 1920s in the United States saw its own unique climate of emotional control, inspired in large part by the new technologies of the era. Just as these technologies were met with conflicted reactions, this new emotional climate was complicated and contradictory, highlighting the uneven ways that scientists, teachers, businesspeople, and others dealt with the ideas about emotion that developed with the new media age.

The speech discipline's movement away from elocution offers an especially strong example of the emotional transition to American cool. As Herman Cohen has argued, during the nineteenth and early twentieth centuries, "elocution was the dominant means of teaching oral expression in American colleges and universities."[3] In addition to its academic popularity, Cohen explains, "elocution became deeply imbedded in the culture of the time. In a literary-oral society Elocution became an important form of entertainment and, even, of literary improvement."[4] At the end of the nineteenth century, however, this highly cultivated, literary style of speaking had begun losing its cultural force. According to Kenneth Cmiel, "by the turn of the century, technical, plain, and colloquial styles were all presented as alternatives to traditional rhetorical ideas about speech."[5] Cmiel emphasizes how such developments as populism and the democratization of education helped to move elocution out of the popular and academic spotlight. In the early years of the twentieth century, Michael Leff and Margaret Procario explain, "academics had come to regard elocution as a cosmetic technique totally devoid of substance."[6] The growing culture of emotional control is one reason for this academic and popular rejection of elocutionary practice.

As I discuss in chapter 1, elocution was a highly theatrical, highly emotive style of speaking. Stressing the centrality of emotion to elocution, a book by John Walker first published in 1781 emphasized the importance of the "plaintive" speech that would become the target of many twentieth-century critics:

> The noble and generous passions are the constant topicks of ancient and modern poets; and of these passions, the pathetick seems the favourite and most endearing theme. Those readers, therefore, who cannot assume a plaintive tone of voice, will never succeed in reading poetry; and those who have this power, will read verse very agreeably, though almost every other requisite for delivery be wanting.[7]

For Walker, this powerful emotionality needed to carry over from one's voice to one's gestures. Although Walker believed that speakers should "be sparing in the use of the left hand," he offered elaborate instruction on how speakers should use the right hand to demonstrate a sufficient level of passion. Walker argued that the speaker's right hand "ought to rise extending from the side, that is, in a direction from left to right; and then be propelled forwards, with the fingers open, and easily and differently curve." He would add that "above all, we must be careful to let the stroke of the hand, which marks force, or emphasis, keep exact time with the force of pronunciation."[8]

While these earlier thinkers on oratory understood that such prescriptions might seem artificial, they held to their inherent emotional power nonetheless. Against the charge that elocutionary speech might be unnatural, J. W. Shoemaker argued that "it is necessary to assist Nature by careful cultivation in all that pertains to Expression." She would go on to explain that "mind and spirit communicate themselves rapidly and often passionately to the outer world through the body medium, in ways which may be natural, but which are by no means perfect or graceful expressions of Nature."[9] For instance, although someone's natural inclination might be to pronounce an r in a smooth manner, for truly emotive oratory, Shoemaker explained, when preceding a vowel the r should be trilled. "The degree of the trill is governed by the character of the sentiment," she elaborated. "In bold, impassioned utterance, and in all forms of dignified discourse, the trill should be quite decided."[10]

Alongside the new climate of emotional restraint identified by Stearns, this more "impassioned" mode of speech gave way to a different model. As I discuss in chapter 1, the presumed chaos of the new technological age was seen to have a whole host of negative effects, including the speed mania associated with the automobile and the more general sense of bodily shock attributed not only to the war but to the wider stresses of the modern age. In a similar way, the emotions that were once seen as a central component of a powerful, elocutionary speaking style came more and more to be seen as bodily disturbances that needed to be controlled in order for the voice to function properly. A 1915 essay in the *Quarterly Journal of Public Speaking* by Smiley Blanton attempted to clarify this relationship between emotion and the physiological conditions of speech. When a person experiences pleasurable emotions such as joy and love, Blanton argued, "eyes brighten, cheeks redden, tense muscles become relaxed, wrinkled brows smooth; the voice becomes soft and more pleasing." "All this," he clarified, "refers to mild, controlled, pleasant emotions." In contrast, "extreme emotions of any kind" had an opposite effect on speech: "the energy of the body is used up; digestion is halted; breathing becomes irregular and usually more shallow; the voice changes."[11]

In a public speaking textbook, Kathleen O'Keeffe similarly warned that "there is of course a grave danger involved in development of the emotions." O'Keeffe drew on the same James-Lange conception of emotions that Albert Osborne had associated with the stereoscope: "Dr. James has pointed out that every emotion aroused must find a channel of expression through action, or it will be thrown back on the consciousness and a complex started."[12] A 1916 essay by F. H. Lane argued that because of the speech discipline's movement toward emotional constraint, American public speaking had begun to reach a higher level in regard to the treatment of emotion. "By the application of scientific methods in the treatment of their themes," Lane asserted, "speakers have been able to advance the suffrage and prohibition movements much more rapidly than the speakers who relied upon emotional appeal."[13] For Blanton, O'Keeffe, and Lane, control of a speaker's emotions led to more controlled bodies, pleasing speech, and persuasive arguments.

The 1910s and 1920s saw elocution largely abandoned in favor of the new public speaking. In 1922, when the speech- and theater-centered Emerson College hired John Connor as its first professor with public speaking in his title, the college's three professors of elocution each had their title changed. Priscilla Puffer, who had been listed as professor of gesture and elocution in 1921, was now listed as professor of gesture and expression. Margarette Penick's areas of study were changed from elocution and recitals to lyceum and Chautauqua programs. Francis Joseph McCabe became a professor of dramatic interpretation.[14]

In 1923, Wayland Parrish lamented the remaining vestiges of elocutionary practice, claiming that "in spite of the modern drift toward a more practical style of speaking, students are still coming to college and, alas, leaving college, with the notion that a speech to be excellent as a speech must be composed in what Paul Shorey calls 'the florid, antithetic, jingling style of sophomoric ornament.'"[15] Dale Carnegie had just as little respect for elocution, writing in a 1926 textbook that "an enormous amount of nonsense and twaddle has been written about delivery. . . . Old-fashioned 'elocution,' that abomination in the sight of God and man, has often made it ridiculous."[16] By 1930, Edward Rowell of the University of California argued that "professors of Public Speaking appear to be unanimous in urging that the proper mode for speakers in our day is the conversational as against the elocutionary or the formally oratorical style."[17] In 1933,

Lawrence Goodrich felt confident enough in the "new public speaking" to write that "to create the illusion of real talk should be a goal in all speech arts."[18] The florid, plaintive style of elocution had been pushed aside by a more presumably conversational but emotionally controlled one.

This rejection of elocution in favor of the new public speaking was widespread in the discipline of speech, even among scholars who otherwise seemed to disagree with each other. Leff and Procario suggest that the speech tradition that emerged in the early twentieth century was dominated by two different schools of thought: an emerging scientific school, which believed speech should be studied through rigorous scientific method, and a "Cornell University School," based in the humanities, which emphasized traditional rhetorical ideas.[19] As much as Leff and Procario attempt to differentiate these traditions, however, they largely shared the new emphasis on practical, emotionally controlled speech. James Winans, whom Leff and Procario identify with the scientific side of this division, was at Cornell for a period of time in which he developed courses in the new public speaking. As Winans explained in an essay describing the basic public speaking course at Cornell, "our students are impatient of the niceties of elocution and rhetoric; and in the limited time they give to our work they would profit little if at all by them." Instead, "the aim of the course is practical public speaking."[20] Similarly, in a discussion of the eighteenth-century orator Joseph Priestly, Hoyt Hudson, whom Leff and Procario identify as a humanistic scholar, celebrated this earlier speaker's ability to perform the conversational speech overwhelmingly endorsed in the 1920s. "For our present purpose," Hudson claimed, "the best summary is the comment of one who had often heard him preach: 'He uses no action, no declamation, but his voice and manner are those of one friend speaking to another.'"[21] Humanistic and scientific speech scholars alike saw value in the more practical, emotionally controlled, conversational model of public speaking.

This shared rejection of the more emotive elocution for the new model of public speaking developed in parallel with the technological aesthetic of the voice that emerged during the 1920s. Although not all speech scholars made use of the new recording technologies, the pervasiveness of these technologies, coupled with growing concerns about emotional control that were influenced in part by their presence, encouraged a series of changes in speech practice and pedagogy. Exploring how these technologies were utilized in speech research illustrates some important ways this technological aesthetic took shape and thus offers a key to understanding the larger emergence of the new public speaking.

Technology and the Science of Speech

The emotional and technological climate of the 1920s had profound impacts on academic research. During this period, fields such as psychology and sociology tried to gain academic legitimacy by aligning themselves with the natural sciences and pushing beyond their more emotionally tainted pasts. A growing number of psychologists aimed to create a separation between themselves and the philosophers and spiritualists with whom they shared a common history. They also worked to distance themselves from the method of introspection that had been popular in the nineteenth and early twentieth centuries—a movement explored in more detail in the next chapter. Here, psychologists had asked their subjects to reflect on their own experiences in ways that were apparently too touchy-feely for the new climate of emotional control.[22] Sociologists' shared history with social workers proved a similar problem.[23]

Because of its historical connections to elocution, the study of speech had a correspondingly emotional background to shed. Arguing for a less emotional form of speech was one step in this direction. Another step was the adoption of the tools and techniques of the natural scientist. As was the case with various psychologists and sociologists of this period, many studies of speech began to use a range of recording devices intended to objectively measure research phenomena. Devices such as psycho-galvanometers had been popularized in the nineteenth century by Wilhelm Wundt and other psychologists, and by the 1930s their use was widespread throughout the social sciences.[24] These technologies provided the veneer of scientific validity and objectivity demanded by a culture concerned with emotional control. Researchers could distance themselves from their own emotions at the same time that they measured, analyzed, and controlled those of their subjects. The range of technologies applied to speech served to reinforce the ideal of emotional control at the same time that they advocated a new technological aesthetic of the voice.

Edward Wheeler Scripture, who had discussed the psychological importance of stereoscopic technology, was one important advocate of using recording technologies in the study of the voice. In a 1901 essay in *Modern Language Notes*, Scripture described an apparatus built in his Yale psychological lab that created sound "tracings" from phonograph records. The machine created a horizontal graph of a record's vibrations and thus offered a visual representation of its tones, pauses, and other auditory data. Scripture argued that these tracings allowed for a very detailed examination of the qualities of different recorded sounds. Although "a careful study of the sound by the ear reveals some of the grosser characters of the sound," Scripture contended, simply listening to a sound could not "indicate any of the finer details that lie before the eye in the complexities of the curve." From Scripture's perspective, there were important elements of a sound that went unnoticed by the unaided ear.[25]

In a 1902 essay in the *Century Illustrated Magazine* entitled "How the Voice Looks," Scripture described his voice research for a more popular audience. Here, Scripture again stressed the centrality of vocal vibrations to the power of speech. According to Scripture, "the voice issuing from a person's mouth consists of vibrations of the particles of air; these vibrations represent the entire effect of thought and emotion that pass from the speaker to the hearer." As a result, Scripture argued, his visual tracings of speech curves provided a very powerful means of understanding the vocal expression of emotion:

> We can all detect sorrow, anger, fear, fatigue, etc., in a person's voice. Since the voice travels in the form of air-vibrations from the speaker's mouth to the ear, a record of these vibrations must contain the results of emotions. An understanding of the modifications of the speech curves must reveal the effects due to the emotions.

Owing to the technological power of these recordings, Scripture concluded, studying one's own and others' speech curves could give a much stronger understanding of vocal emotions.[26] The emotions of speech were but a collection of vibrations flowing from speaker to listener.

Another research technology developed in Scripture's lab was the "sound cage." The first such apparatus had been designed by Matataro Matsumoto while performing sound research under Scripture. It consisted of a spherical metal cage suspended over a chair. A subject would be placed in the chair and then sounds would be produced at various points in the sphere as a test of the subject's ability to judge the direction of these sounds. Suggesting his faith in the technologies of the period, in the article where Matsumoto introduced this device, he included a stereoscopic diagram as a means of demonstrating—in the presumed three dimensions of a high-tech stereoscope—the different axes of sound production around the surface of the sphere.[27] Matsumoto's sound cage was championed by such established scholars as Cornell University's Edward Bradford Titchener, whose psychological laboratory employed its own array of sound-related instruments. In addition to a sound cage, Titchener's lab included an electric phonograph, a xylophone, and a number of metronomes, pianos, harmonicas, ocarinas, and other tools for producing and measuring sounds.[28]

Carl Seashore, one of Scripture's most prolific students, developed an elaborate program in technology-centered analyses of sound in the University of Iowa's psychological lab.[29] In an essay published in 1902, Seashore described what he termed a "voice tonoscope." The device created a graphic representation of the sound waves produced by vocal performances; "the vibrations of the voice [were] made visible upon a moving surface by the action of intermittent light."[30] Visualizing sound waves in this manner, Seashore echoed Scripture, allowed for an objective, scientific understanding of the voice. He went so far as to suggest that his graphs of vocal performances provided a better appreciation of the beauty of sound than was possible for even a highly trained listener. "The photographic reproduction of the sound has a far more faithful detail than even the most musical ear can hear," he argued.[31] The true beauty of the voice lay below its surface, in the technical details of the sound wave.

In Seashore's efforts to objectify the emotional elements of sound, even beauty became a statistical product. Beautiful vocal performances resulted from a pleasing "deviation from the regular." When untrained voices varied from the regular, the result was typically ugly because "the artist who is to vary effectively from the exact must know the exact and must have mastered its attainment before his emotion can express itself adequately through a sort of flirtation with it."[32] Seashore's technologically driven sound research aimed to quantify the range of deviations that created this objectively defined beauty. In a study employing "phonophotography"—an enhanced version of his tonoscope—to capture and compare the sound waves produced by a series of singers, Seashore

Figure 3.1
Tonoscope. Reprinted from Carl E. Seashore, "A Voice Tonoscope," *University of Iowa Studies in Psychology* 3 (1902): 21.

quantified the range of variations that made for a beautiful vibrato. The
vibratos of the best singers, Seashore reported, oscillated at an average
of a half-tone and "at an average rate of six or seven cycles per second."[33]
Through such analyses, Seashore hoped to create an objective measure
of musical artistry that could be passed on to musicians as a scientific
means of improving their performances. He also established a normative level of emotional expression, offering a very concrete, technical
argument in support of a perfectly average emotionality.

Glenn Merry, a colleague of Seashore's in the speech department
at the University of Iowa, used these same technological measures to
record and analyze speech, creating vocal tracings much like those of
Edward Scripture. For one early study, Merry used phonograph recordings in an attempt to research the links between speech and nasal
resonance. Merry aimed to learn the extent to which good speech

was connected with more or less vocal resonance in one's nasal passages. Using the students in an introductory speech class, Merry and his fellow teachers picked the fifty best and fifty worst speakers from among a group of classes—going for the kind of average or cross section that Seashore had explored. Merry and his research assistants then recorded these hundred students as they spoke into the horn of an "Edison Opera Talking Machine." Once the students' voices had been recorded, the researchers replaced the phonograph machine's recording horn with rubber tubes, which were inserted in the students' nasal passages. This time, as the students spoke, the phonograph record recorded the sounds coming directly from their noses. Based on his data, Merry concluded that "it seems from the study that the voices of both men and women may predominate either in strong, medium, or weak nasal resonance."[34]

In another essay published in a journal issue edited by Seashore, Merry aimed "to develop a method for determining objectively the pitch of the human voice, in any or all of its inflections in speech."[35] Merry had attached a tonoscope to a record player so he could chart the pitch variations in a series of speeches that included Franklin Roosevelt's "Americanism" and one of Portia's soliloquies from *The Merchant of Venice*, recited by Julia Marlowe. For each of the speeches analyzed, Merry produced a diagram that traced its pitch variation on the chromatic scale. In the line "much has been said of late about good Americanism," for instance, Roosevelt's pitch apparently rose more than an octave, from a low A to just above middle C. In this second study, in order to follow Seashore's model, Merry had first turned speech into music and then used the tonoscope to offer a presumably objective graphic representation of it. For both Merry and Seashore, the artistic and emotional elements of the voice had been reduced to the tracings of their laboratory devices.

Merry also encouraged the use of X-rays, another progressively more popular technology for measuring and analyzing speech. In an essay for the *Quarterly Journal of Speech Education* promoting a "roentgenological method" of analysis, he argued that X-rays (roentgenograms) could provide a more accurate understanding of voice "placement." Merry instructed researchers to "place the subject before the fluorescent screen and let him speak or sing while turning his body so that

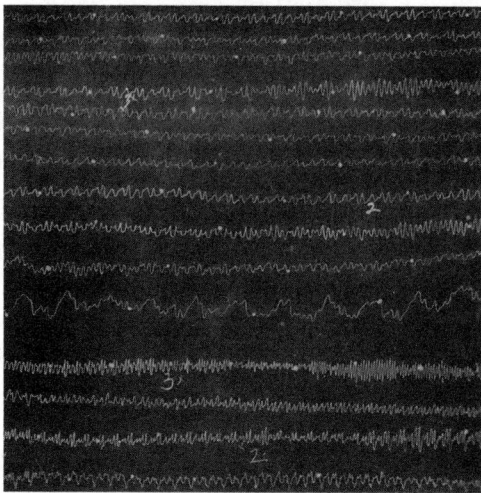

Figure 3.2

An image of a speech wave form. Reprinted from Glenn N. Merry, "Voice Inflection in Speech," *Psychological Monographs* 31, no. 1 (1922): 208.

the roentgenogram shows both postern-anterior and lateral views." Having done so, he explained, "the adjustment of the organs *regulating resonance* is plainly visible."[36] In another essay, Merry reported on a case of a young female student at Iowa who had taken speech classes "for two and a half years with very little improvement in a voice quality that was decidedly unpleasant." After completing an X-ray of the woman's sinuses, Merry explained, "a radiograph showed that no amount of training would ever give her a good voice. The chambers of resonance above the palates were shallow and narrow."[37] In Merry's analysis, the more traditional methods of speech instruction had been no match for the technological and biological truth of the X-ray. In establishing an average size and depth of the sinuses, Merry believed that he had established a biological baseline for good speaking ability.

G. Oscar Russell's 1928 book *The Vowel* used X-rays as a means of understanding vowel positioning in the mouth. The book includes a number of X-ray images that serve to support Russell's arguments. The caption from a typical image illustrates the form of speech Russell strove to document:

Vowel *i* (peep) Mid-East American. Male. Cultured but non-pedantic pronunciation. Normal un-impeded speech.[38]

This idea of "cultured but non-pedantic" speech—much like Seashore's average beauty—played an important role in these scholars' thinking about the voice. As far as Russell was concerned, he could get an objective picture of typical—and cultured—speech through the apparatus of the X-ray machine.

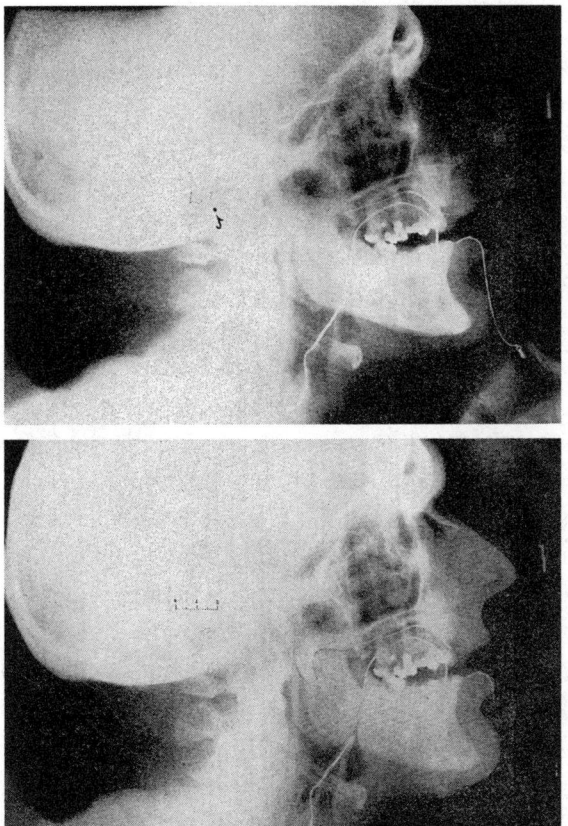

Figure 3.3

X-ray of mid-East American male pronouncing the vowel *i* (peep). Reprinted from G. Oscar Russell, *The Vowel: Its Physiological Mechanisms as Shown by X-Ray* (Columbus: Ohio State University Press, 1928), 257.

In its expert scientific vision, the X-ray served Merry and Russell much as the tonoscope did Seashore. For all three researchers, to appreciate the emotional normalcy of the speech they explored, the scientist needed to get below the surface, where only these technologies could go. Like Seashore, both Russell and Merry believed that their technologies gave a truer understanding of vocal performance than had been available by earlier means. Merry claimed to have disproven the taken-for-granted idea that vocal performance took place "in the head," and Russell challenged the standard notion of vowel positions. The "deeper" these technologies allowed researchers to see within their subject, the deeper the true or natural voice seemed to be.

In addition to these recording and X-ray technologies, speech researchers were also experimenting with radio technology. In one *Quarterly Journal of Speech* essay, Claude Merton Wise of Louisiana State University reported on a study that used elaborate radio transmitting equipment to study the "chest resonance" of speakers. In order to hear the sounds coming from the test subjects' chests, Wise and his fellow experimenters created "a highly complicated assembly of radio units, beginning with a broadcast microphone and ending with a dial calibrated in decibels to record the slightest fluctuations of the volume of sound."[39] Wise reported that the "high sensitivity" of this noise meter allowed the experimenters to differentiate between minute elements of chest resonance that earlier speech researchers had confused. As with the X-ray and tonoscope, the sophisticated technology of radio amplification promised a purer access to the essential elements of speech.

Like Merry's speech department at the University of Iowa, the University of Wisconsin's department of speech was doing its own technology-centered explorations. As was the case with many other speech programs, Wisconsin's speech coursework was in the English department at the beginning of the twentieth century. As of 1906, the catalog for Wisconsin's English department listed courses in rhetoric and oratory, which included multiple classes in elocution and declamation in line with the more traditional nineteenth-century program in elocution. By 1907, this area of the English department had changed to "Rhetoric and Composition," and the university's courses in speech had been spun off into a different program.[40]

The new Wisconsin program in speech emphasized the more presumably practical speech that made up the developing public speaking tradition. In line with this, the program initially adopted the title "Public Speaking" (which was changed simply to "Speech" in the early 1920s). Wisconsin's department drew very clear connections between this new model of speech and the increasingly technological research that was becoming prominent among speech scholars such as Merry. The emotional-control advocate Smiley Blanton was appointed to Wisconsin's department in 1914. When the then department chair, James O'Neill, recommended Blanton to his dean, he stressed both Blanton's background in speech education and his recently completed medical degree as fundamental to his qualifications. "Mr. Blanton has taken his medical course in addition to his special training in speaking for the purpose of fitting himself for work in correcting speech defects and developing proper vocal methods in students," O'Neill wrote.[41] Blanton's medical degree was seen as an important addition to the department because, like Merry, Wisconsin's department viewed public speaking as a largely technical practice. Like the apparatuses through which speech could be recorded and analyzed, the human voice was its own sort of technical apparatus that required the minute adjustments noticeable by X-ray—or nasal tube—and through the physiological expertise of a doctor.

As did Seashore, Scripture, Merry, Russell, and Wise, Wisconsin's speech department employed a number of different apparatuses in its studies of speech. Showing his interest in physiology, Blanton suggested that speech scholars use "laryngoscopes" to study the anatomy of vocal production, and even offered that sheep larynxes could be a valuable part of such research. Recognizing the importance of recording technologies, he also promoted the use of rotating recording drums similar to the apparatuses that Seashore and Merry had employed.[42] Blanton's Wisconsin colleague Andrew Weaver likewise used an impressive array of recording technologies. In 1919, Weaver wrote a letter to Thomas A. Edison Inc., requesting a "phonograph which will record and reproduce the speaking voice as accurately as possible."[43] Weaver also ordered an artificial larynx from Western Electric's division of scientific instruments, a model thorax from Denoyer-Geppert Company, and an audiometer from Graybar Electric via Western Electric's Bell Labs.[44]

Like Merry, Weaver believed that the application of scientific equipment to speech would provide a more objective, scientific basis for speech practice that would improve the field as a whole. Weaver suggested that

many nonscientific studies of speech—including those of the earlier elocutionists—imagined that "the voice is an indissoluble part of a mystical, occult, transcendental entity usually designated as *personality* or *soul*
which is by nature above and beyond analysis."[45] Weaver believed that the
scientific apparatuses he used in his research offered a way of overcoming this flawed perspective. Using a "phonautograph," Weaver produced
graphic tracings of recorded speeches similar to that created and analyzed
by Scripture and Merry, both of whose work he drew upon. Relying on
his phonautographic measurements, Weaver postulated a series of averages regarding male and female vocal pitch, vocal inflection, and pitch
memory, establishing a set of rules of thumb for normal speech practice
similar to those put forward by Merry and Seashore.

One of Wisconsin's most ardent promoters of these technologies was
Robert West. West joined Wisconsin's speech department as a master's
student in 1918. He earned his MA in 1920 and then continued on for his
PhD under the advisement of Smiley Blanton, working briefly in Iowa's
speech department in the interim. Eventually he joined Wisconsin's
department as a professor.[46] As both a graduate student and professor,
West utilized the full range of technologies available to speech researchers. In a 1924 discussion of the "telegraphone," a wire recording apparatus
that West had experimented with while teaching speech classes at Wisconsin, West declared the machine "one of the most significant mechanical devices that have been produced in modern times to help in the
training of public speakers."[47] This device allowed West and other speech
instructors to record students' speeches for future playback and analysis.

A year later, West described how the same device—which he now
identified as a magnetophone—could be used to chart and analyze
speech in much the way that Seashore and Merry had done with the tonoscope. Because the recorded sounds left a magnetic trace on the wire
recording apparatus, West discerned that an image created from the wire
would offer an accurate picture of a given vocal performance. Since there
was "only one mechanical process between the speaking and the analysis
by the compass needle" and no amplifying bulbs were used, West argued
that there was little opportunity for the signal to distort. As a result, West
concluded, "the magnetic arrangement of the molecules in the steel wire
is a fairly accurate picture of the sound waves that are being studied."[48] As
was the case with Seashore, Merry, and Russell, for West, the recording

capabilities of the magnetophone modeled a kind of technological truthfulness. By subverting the distortions of the human ear, eye, and common sense, these recording apparatuses pushed away the ornamental surface of sounds in favor of their deeper acoustical or physiological reality.

West's dissertation, which he published a year after his magnetophone study, brought his technological thinking full circle. Although he continued to pursue his magnetophone research, West also built his own artificial larynx and used it to model and analyze vocal production. He stretched rubber across a piece of tubing and then slit it to produce two vibrating edges. Having done so, he "could send a blast of air through the device and watch its effect upon the edges of the slit in the rubber tympan."[49] Using his magnetophone, West would "tune" the mechanical larynx so that it approximated the actual sound waves that he saw being produced by the human voice. West then observed

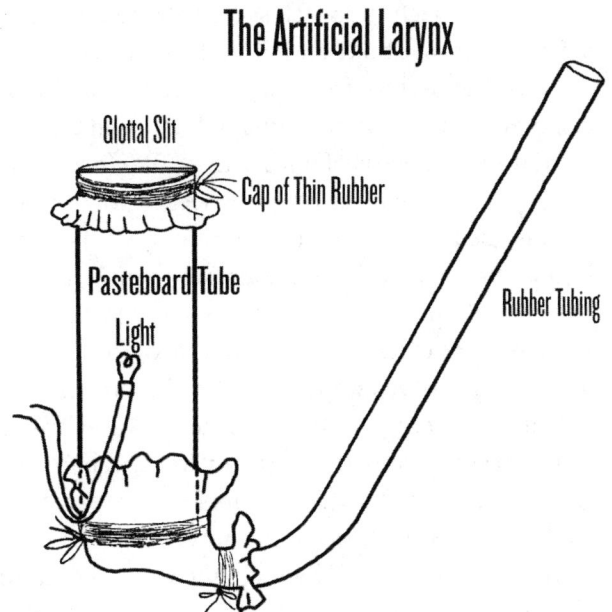

The Artificial Larynx

Glottal Slit

Cap of Thin Rubber

Pasteboard Tube

Light

Rubber Tubing

Figure 3.4

Diagram of artificial larynx. Reprinted from Robert West, "The Nature of Vocal Sounds," *Quarterly Journal of Speech Education* 7, no. 4 (1926): 284.

the artificial larynx in action, using his observations to offer arguments about the physics and physiology of the actual human larynx.

West's research exemplifies the thinking in much of the wider tradition of technology-centered speech analysis on which he drew. For the scholars involved in this research, these recording and other voice technologies understood a deeper truth about the voice than could be gleaned from merely listening to someone speak or sing. These technologies were seen as better listeners in large part because human beings themselves had come to be imagined as technologies; people were collections of physiological mechanisms that, if tuned correctly, could produce beautiful speech. West's research using his artificial larynx illustrated this quite clearly. In this research, the "human" component of speaking existed merely as a series of technological traces flowing back and forth from the rubber tympan to the magnetophone. For West and these other speech scholars, the best speech would emulate the recording technologies of the laboratory, and those technologies would themselves be the best judge of when this beautiful speech had been achieved.

This technological approach to speech served the new public speaking in several ways. For one, it gave both speech students and speech researchers a way to distance themselves from the hyperemotional practices of the elocutionists. In order to speak well, one did not need to call up some powerful emotion from the depths of one's soul, as the elocutionists of the previous century had suggested. Rather, one needed to be a scientist of the voice, which was itself a special sort of technology. This technological take on speech lined up well with the presumably plain style of the new speech, in that the attention to internal, physiological techniques further de-emphasized the sorts of ornamentations that had occupied the elocutionists. The emotions of a speech were not located in some elaborate hand gesture or vocal trill; they were to be found in sound waves or the specific movements of the larynx. From this perspective, speaking "naturally" meant employing the kind of statistically supported, average intonation that Seashore had associated with beautiful music. In a like way, in their presumed objectivity, the new technologies of the voice placed the speech researcher on a level with the natural and medical scientist, further highlighting the rational, scientific legitimacy of speech research and practice.

In using recording technologies to escape a perceived problem of emotions, this research tapped into a still longer history of sound recording, as Jonathan Sterne has demonstrated. Seashore's tonoscope was a descendent of the manometric flame that Alexander Graham Bell had used in his experiments with phonography and deafness. While Bell framed himself as a champion of the deaf cause, however, others ultimately saw him as attempting to eradicate deafness in the same way that eugenicists attempted to get rid of other presumably genetic problems. For this reason, writes Sterne, sound recording "arose, in part, from an attempt among hearing people to 'solve' the cultural problem of deafness."[50] The early twentieth-century work of Seashore and other speech researchers arose amid concerns for controlling the dangers of emotional stimulation of the voice and body. In locating the emotions of speech in a series of laboratory apparatuses, the new speech seemed to free both speaker and scholar alike from the emotional disturbances of the new media age.

Selling Scientific Sounds

In addition to the concrete changes to speech research and practice brought about by this new technological emphasis, the growing dependence on these technologies also placed speech scholars in close contact with a range of equipment manufacturers. Andrew Weaver's correspondence with the Edison Company was but one example of these interactions, as speech researchers became important consumers of various commercially available recording instruments. However, the on-the-ground experimentation with this equipment required a number of modifications, and many of these researchers quickly became experts on the production of these instruments as well. As a result, a number of researchers turned their attention to designing and marketing their own speech-related technologies, which were sold to other researchers, schools, and even the general public. In the process, these speech scholars publicized the technological aesthetic of the voice lab still more strongly, highlighting the essentially technological nature of good speech and selling a vision of technological and emotional control to the public at large.

Chicago's C. H. Stoelting Company, one of the most prominent producers of scientific technology in the early twentieth century, was active in distributing research tools created in speech and other research labs. The company's founder, Christian Hans Stoelting, had earned a patent for an "autographic recording device" in 1894. Stoelting and his then business partner, Robert Copeland, described the device as useful for "making written memoranda, such as shipping manifests, bills of lading, cash or sales checks, and various other kinds of memoranda."[51] A year later, Stoelting received a patent for a pocket camera that held a roll of film inside.[52] At the same time that he was producing these devices for wider public use, Stoelting was beginning to manufacture the scientific instruments that would find an important place in early twentieth-century research labs. He worked closely with Titchener's Cornell lab from its founding and also went on to earn patents for a barometer and a set of laboratory weights.[53]

If Stoelting had begun as a manufacturer for popular audiences before turning to the lab, Carl Seashore and a number of other sound and speech researchers had begun producing equipment for their labs and then made them available to wider audiences. An audiometer that Seashore designed for his research was manufactured and distributed by Stoelting beginning in 1900.[54] Drawing on his research on the psychology of music, beginning in the 1910s, Seashore also produced a test of musical talent that was available to secondary schools and others throughout the country. The test's manual of instruction—also produced and distributed by Stoelting—opened on a celebratory note:

> Just as the great musicians live before us now in the wonderful reproduction of the modern phonograph, so the intricate experiments of the psychological laboratory may now be popularized by the faithful reproduction of the sounds of laboratory instruments and their scientific presentation.

The test included the manual and a set of phonograph records designed to test people's sense of pitch, rhythm, tonal memory, and related musical abilities. A test taker would listen to a set of tones or rhythmic beeps, and then answer a set of questions on the included test booklets. Based on the psychological research that supported it, the test's manual promised its users "quantitative results which may be verified to a high degree of certainty."[55]

Seashore's music tests found their most eager audiences in schools throughout the country—even among such established programs as the Eastman School of Music.[56] According to Seashore's instructions, secondary students were to be tested in fifth grade and then again in eighth grade so that those with exceptional musical ability could be singled out for additional instruction and cultivation. Seashore suggested that the tests could also be used as material for experiments in elementary psychology classes and would "also fill a great need in the theoretical instruction in the music school." In addition, Seashore explained, the tests "furnish also material for scientific entertainment in the home. Taking one test each evening, this outfit provides material for six evenings of delightful entertainment in the form of a competitive game."[57] Seashore's psychological experiments with sound had been turned into a presumably scientific set of records that promised entertainment and education far beyond the walls of the laboratory.

Smiley Blanton was also working with C. H. Stoelting. Together with his wife, Margaret Blanton, and Sara Stinchfield, both graduate students in Wisconsin's speech department, Blanton created and sold the "Blanton-Stinchfield Speech Measurements." Like Seashore's tests of musical talent, the Blanton-Stinchfield tests were designed to measure a person's talent for speaking. The test was to be given to students from preschool and kindergarten through eighth grade, with an additional scoring scheme for adults.

Figure 3.5

Phonograph record for Carl Seashore's Test of Musical Talent. University of Iowa Libraries. Photograph by author.

A set of "subjective measures" offered a score of a person's "speech reaction manifested in behavior, emotional type, specialized muscle movements, postural tensions, physical anomalies; vocal quality, pitch and volume, respiration and speech defect, if any." A second set of "objective measures" provided measurements of a person's articulation, oral and silent reading rates, spontaneous speech rate, use of relevant words, and vocabulary.[58]

In a portion of the exam intended for preschool and kindergarten students, the student would be shown a picture and then have his or her response transcribed. The response would then receive a score of "superior," "average," or "poor or unsatisfactory," based on the student's articulation, rate of speech, and so forth. These scores were based on a template provided as part of the test book. For instance, after being shown a picture of "this little pig," a superior response would entail a well-articulated, word-for-word recitation of the rhyme "This Little Pig Went to Market" (though the test's authors offered that "wost" beef could be accepted in place of "roast" beef, given the age of the test takers). An average response would entail a basic—but clearly articulated—recollection of the rhyme, without the word-for-word recitation. An unsatisfactory response would be characterized by "idioglossia," an example of which the test manual transcribes:

> Di li'uh pí di dō to mar'tet
> Di li'uh pí di hā ă hōe
> Di li'uh pí di hă wō bēē
> Di li'uh pí di hă nŭn,
> Di li'uh pí "wēē wēē" aw hōe.

The authors had done their best to offer a scientific rendering of baby talk.

The manual provided for several other concrete speech measurements. For instance, test givers were to calculate the words per minute for each response. According to the Blanton-Stinchfield manual,

> The rate for spontaneous speech varies from 108 to 150 words per minute, in grades one to eight. Platform speakers frequently cultivate a manner of speaking which allows but 75 words per minute. In ordinary conversation our speed more nearly approximates 100 to 150 words per minute. Radio broadcasters speak between 120 and 140 words per minute in making ordinary announcements.

Finally, and in keeping with the technological perspective of Wisconsin's speech department, the manual's authors suggested that a "spirometer" be used to measure a speaker's respiration. This would provide a calculation of a person's "vital capacity," which should average "3600 cc" for eighteen-year-old men, and "2400 cc" for eighteen-year-old women. Like Seashore's music test, the Blanton-Stinchfield test sold a set of technologically produced averages against which various speakers were to be evaluated.

The example of the "Pronunciphone" illustrates a still more commercial way that these technological and scientific understandings of speech entered into the broader culture. The Pronunciphone was a system of phonograph records developed in the 1920s by Edward Hall Gardner and E. Ray Skinner of the University of Wisconsin that promised to teach people to pronounce words correctly. Skinner spent a summer as a student in Wisconsin's speech department in 1921 and then joined the department as a full-time graduate student in 1924.[59] While at Wisconsin, Skinner studied under Andrew Weaver, using audiometers and other such technologies to produce, measure, and analyze various elements of speech.[60] He went on to study such questions as the relationship between pitch and the vocal expression of happiness and sadness.[61] Another member of Wisconsin's department when Skinner was a student remembered him as especially skilled in affecting the diction of the East Coast of the United States—one of the many abilities that he and his fellow students practiced in their phonetics courses in the department.[62]

Edward Hall Gardner had been an instructor in Wisconsin's English department beginning in 1910, but he gradually began teaching in the department of commerce (later renamed business). Shortly after he arrived at the university, Gardner began teaching a one-semester class on business writing housed in the English department but primarily serving commerce students. Soon after it began, the department of commerce requested that the class be extended to a full year because of its wide popularity. By 1919, the class was housed exclusively in commerce, and Gardner's appointment was split between both departments. In 1921, the school newspaper reported that Gardner's class had broken university-wide records for enrollment, with 420 students enrolled.[63]

In addition to his business writing class, in 1915 Gardner published a book entitled *Effective Business Letters*. Here, he advocated a style of writing that was both formal and personal, attempting to temper the kind of sentimentalism that had characterized much nineteenth-century letter writing while still encouraging writers to express their personalities. According to Gardner, when writing a business letter, one should "imitate the tone of conversation," writing "as cordially and personally as if you were face to face with your correspondent." However, "letters must always be more dignified than conversation: . . . just as it is bad taste for a salesman, in matters of business, to act with all the informality of a friend, so it is bad taste for business letters to copy exactly the appearance and the style of social letters."[64] Mastering good business communication was central to being successful in commerce. "Most business men realize that letters perform nine-tenths of the work of business, and that consequently better letters are as necessary as better cost-keeping or better sales methods," Gardner asserted.[65] Good business writing elevated one's capital just as these other business improvements could.

Unlike Seashore and Blanton, Blanton, and Stinchfield, who relied on C. H. Stoelting to distribute their tests of talent, Gardner and Skinner started their own business, the Pronunciphone Company. The Pronunciphone record system reflected the combination of Gardner's idea of language as a kind of capital and Skinner's scientific approach to the voice. Poor pronunciation, the company's ads warned, would cause personal embarrassment and decrease one's personal and business capital. "Are You Embarrassed by Mistakes in Pronunciation?" read the headline for one advertisement; "Nothing reveals your culture—or lack of it—so surely."[66] Another ad promised "CULTURED SPEECH—by a new method."[67] These ads explained that improper pronunciation is "A Serious Social and Business Handicap," and suggested that if you had not yet developed the kind of cultured speech they promised, "you are cheating yourself of a tremendously effective social and business asset." Clarifying the kind of embarrassment of which these ads warned, another headline read, "'Faux Pas' I said . . . And Everyone Tittered." The ad continued with a first-person narrative regarding the speaker's offending "fox pass" and ended with the triumphant story of her successful use of the Pronunciphone.[68]

As Gardner advocated in his teaching on business writing, the proper speech taught by the Pronunciphone was to be both *cultured* and *conversational*, encompassing both a systematic formality and an everyday casualness. As an advertisement in the magazine *Forum* explained,

Figure 3.6

Pronunciphone advertisement. Reprinted from *English Journal* 20, no. 10 (1931): 876·

In addition to hundreds of words of general use, there are included words used by cultured persons in discussing art, literature, history, biography, science, and geography. There are also many popular foreign words (French, Spanish, Latin, etc.), that are now an essential part of the educated American's vocabulary.

According to Skinner and Gardner, educated people could be expected to pepper their everyday conversation with such words as "Beethoven," "hors d'oeuvres," "impious," "Buenos Aires," "psychiatry," "canapé," and "naïve"—all of which they listed as commonly mispronounced. Other words listed under the caption "How many of these words *dare* you use in conversation?" included the names of the silent film actress Renée Adorée and the Italian tenor Tito Schiapa.[69] A cultured conversationalist was one who could chat about classical music, food, and Hollywood actresses, all without missing a rolled *r*.

In line with Skinner's research on speech, the Pronunciphone system was marketed as a highly technological solution to pronunciation problems. Through the use of "talking machine records electrically recorded in the most modern and scientific manner," the Pronunciphone would

Figure 3.7
Pronunciphone advertisement. Reprinted from *Forum*, May 1928, 8.

provide an expert method for adjusting one's speech.[70] *Good Taste in Speech*, the written manual accompanying the set of records, explained further that "the Pronunciphone records represent what has been declared to be the most difficult feat in the entire history of recording the human voice." This feat had been accomplished through "the use of the latest recording process known in the field of electrical transcription."[71] Companies such as Keystone View and Underwood and Underwood had suggested that the ownership of a stereoscope would not only educate the minds and emotions of consumers, but provide them a certain cultural capital that comes from possessing a high-tech, modern apparatus. Gardner and Skinner attempted to convince consumers that owning and using the Pronunciphone system would elevate their social status through a similar mastering of modern technology.

In this way, the speech practice advocated by Gardner and Skinner's Pronunciphone system drew together the controlled speech of the new public speaking and the technological perspective of the speech lab—all in one popularly available form. If, as Seashore and other researchers suggested, good speech was primarily a technical matter of properly produced sound waves, then speakers could be expected to emulate technologies in order to speak effectively. The Pronunciphone system encouraged its users to do just this; they were to model their speech after a phonograph record, pronouncing words in a predetermined way time and time again. In fact, the very process of using the Pronunciphone cast the speaker as an automaton endlessly playing its phonograph. Listeners were expected to "listen to ten words at a time, then set back the tone arm, repeat[ing] the words as often as desired until [they] had thoroughly mastered them."[72] Only by this endless process of repetition could speakers achieve the ideal balance of cultured conversation that would help them advance both socially and in the business world. Like the writers of Gardner's ideal business letters, users of the Pronunciphone were to speak with a perfect mixture of individual personality and systematic, formalized pronunciation.

Finally, the Pronunciphone, Seashore's music tests, and the Blanton-Stinchfield Speech Measurements all celebrated the sort of "pleasing deviation from the regular" by which Seashore had defined beauty. Seashore's test was based upon his studies of the psychology of music. Using the averages he obtained in his lab, he had created a template for judging a person's musical ability by comparing her or him against others with presumably strong talent. A corresponding test of art talent created by Seashore and Norman Charles Meier provided test takers with two nearly identical paintings and then asked them to identify the better image. Here, too, a scientifically verified average was sufficient to determine which picture was superior and to score the viewer's talent accordingly.[73] Likewise, recognizing that even dictionaries disagreed on the proper pronunciation of various words, for Gardner and Skinner's Pronunciphone system, "the pronunciations of seven dictionaries were compared, and those selected which conformed, as the title of this book indicates, to current good taste in speech."

By crafting an abstracted, scientifically defined understanding of beautiful speech—whether based on appropriate pitch variations or repetitive, culturally approved pronunciations—the technological aesthetic of the voice developed by the Pronunciphone as well as in labs like Seashore's upheld the ethic of emotional control important to both the new public speaking and the speech laboratory that helped to develop it. Musicians, speakers, and scientists themselves were to imitate the tonoscope, X-ray, magnetophone, and Pronunciphone in their ability to appreciate, capture, and re-create the "pleasing deviations from the regular" that amounted to a good vocal performance. By becoming more technological in their approaches, both scientist and performer would presumably achieve the ideal balance of emotional control appropriate for the new media age.

Broadcasting the Cultured Conversation

As the example of the Pronunciphone helps to illustrate, the ideal of speech modeled by the phonograph and other sound devices placed a great pressure on the new public speaker. With the diffusion of recording devices, the public became increasingly aware of the longer-term implications of an individual's speech. Sound had acquired a new sense of permanency, with important implications for reputation and social success. The technological aesthetic that took hold in the 1920s suggested two seemingly contradictory keys to this success. On the one hand, like the tonoscope, speakers were to strip away excess ornamentations and connect with their audience in a natural, presumably unmediated way. At the same time, however, like a phonograph record, they were to speak with a consistency and precision that illustrated an appropriate level of verbal cultivation. They were to be simultaneously conversational and cultured, balancing friendliness with an appropriate level of detached sophistication.

The discussions surrounding the public speech of the radio announcer provide an especially strong example of this developing understanding of the voice and further suggest how these issues stretched beyond the speech laboratory into the larger public. The new technology of radio broadcasting had spawned a range of anxieties about emotion. In 1922 one journalist wrote that "few thoughts which the spiritualists can offer us are more interesting and more 'spooky' than many which are pointed out to those who know of the 'inner workings' of the radio wave. It gives one a chill, for instance, to think that our bodies are constantly acting as conductors of radio waves."[74] Radio waves could also have dangerous effects on the public more generally, as one writer's rant against "communized emotion" indicated: "Several hundred thousand listen to the same speeches every night and the same jazz bands, then at breakfast heave communistic sighs and shed communistic tears over the same calamities, at the bidding of the press."[75] By giving voice to the ether, radio announcers both mitigated and enhanced the eerie, communal power of the radio. The radio announcer engaged in a very public grappling with the new technologies of the broadcast age.

As Paddy Scannell and David Cardiff illustrate in their history of the BBC, the mass distribution of the voice made possible by the radio also created a range of tensions for various self-appointed guardians of the English language:

> Since it was received by family groups it should be conversational in tone rather than declamatory, intimate rather than intimidating. The personality of speakers should shine through their words. But because all broadcasting was live, talks needed to be scripted. Otherwise what they gained in colloquialism and personal idiom they would lose in clarity and succinctness.[76]

These same concerns impacted radio speech in the United States. "Radio, like other mass entertainments, was a site of class tensions and of the pull between homogeneity and diversity," Susan Douglas has explained. "So language use over the air became controversial by the late 1920s."[77] The same balance of friendliness and sophistication celebrated by speech scholars would inflect the speech of the U.S. radio announcer, whose voice both floated through the ether and greeted people in their living rooms.

A U.S. newspaper article from 1924 details the "qualifications necessary to be a radio announcer," noting that "many students of music and elocution apply for radio announcerships." One expert quoted in the article captures the extremes of speech with which radio announcers were supposed to be comfortable, explaining that they "must be able to go from a prohibition banquet to a midnight cabaret and describe each with the same ease and versatility."[78] By 1925, a Radio Voice Technique Committee had formed at the request of radio station WJZ and New York University to begin assessing the ideal radio speech.[79] Among other qualities, the ideal announcer was to speak at 175 words per minute, to introduce marked changes in pace, stress, and pitch, and to "speak in a formal, but friendly, manner."[80] In the committee's second meeting, it rated a selection of announcers on the basis of its recently formed criteria, noting that even the highest-rated person had missed the mark of the ideal radio announcer by a wide margin. On a scale of 100, the highest score was a 66.[81]

Among those tested was Graham McNamee, whom the broadcast historian Erik Barnouw identified as one of the first widely known radio announcers.[82] Although most announcers in radio's early days had been identified through an anonymous sequence of three letters, rather than through their actual name, McNamee had become known for his singing and his recognizable baritone voice. He had moved from Minnesota to New York in order to make it as a singer and then had taken a job at radio station WEAF.[83] His vocal training and background in classical music lent him the air of formality believed necessary for a successful announcer. According to one writer, "Mr. McNamee introduced a touch of culture and refinement, a hint of the Better Things in Life, and a *salon* atmosphere."[84] Alongside his formal speaking abilities, listeners also celebrated McNamee's ability to establish intimate connections with his audience. "It is the universal opinion of hundreds and hundreds of listeners," wrote another McNamee fan, "that the main charm of listening to McNamee is his ability to enter into the spirit of whatever he is reporting and thereby be so wholeheartedly a part of the audience that you too feel you are there."[85] Despite his recognized abilities, McNamee scored a 62 on the radio announcer's exam, placing him four points behind the top scorer.

The ideal radio announcer was to achieve supreme heights of the naturalness and cultivation that were encapsulated in the new public speaking. The same 1924 article that suggested that announcers needed to be comfortable talking about both prohibition banquets and cabarets also provided the following advice for would-be radio speakers from Major J. A. White: "Be Yourself." Another article explained that "listeners dislike unnaturalness and the broadcasters have had to rule out many a promising voice because the candidate was not himself, but, as expressed in a broadcasting phrase, merely 'imitated a broadcaster.'"[86] In a textbook on radio speaking, Glenville Kleiser recounted the story of an early announcer who found success by approaching the microphone "as simply talking to an interested friend," which "guarded him against the common faults of artificiality."[87] Well before so-called reality television, radio announcers may have been the first broadcast entertainers tasked with communicating their own presumably everyday personalities to a mass audience, establishing an early version of the "intimacy at a distance" that continues to be a central feature of broadcasting.[88]

While radio announcers needed to establish a level of intimacy with their audiences, as with the new public speaking more generally, they also needed to demonstrate a level of emotional control. One article from 1926 explaining why male announcers were preferable to female announcers captured the strict boundaries placed on the conversational aspects of early radio speech. Recounting a study undertaken by WJZ, the article argues that women may be less popular as announcers because their voices have "too much personality." Despite admonitions that radio speakers needed to be themselves, according to this article, one of the problems facing female announcers is that their voices are too "highly individual and full of character" to be appreciated by audiences. Repeating cultural stereotypes about women's hyperemotional natures, the article claims that "the listener resents a voice that is too intimate on short acquaintance, and the woman is said to have difficulty in repressing her enthusiasm and in maintaining the necessary reserve and objectivity."[89] The "personality" advocated by these radio experts was of a particularly narrow type.

Other vocal qualities could disqualify someone as an announcer as well. In a *Quarterly Journal of Speech* essay from 1930, Sherman Lawton explained that "a medium low pitch should be striven for in radio speaking for maximum effectiveness. People with high-pitched voices cannot hope to be successful radio speakers."[90] Although announcers were supposed to be themselves, they were also supposed to repress aspects of their voices and personalities that might be off-putting for members of the audience, building intimacy in very narrowly defined ways. The conversational speech of the radio announcer was one that reflected the male, middle-class, white norms encompassed in the dominant emotionology of American cool.

Part of the tension surrounding the naturalness of the announcer's speech, as Scannell and Cardiff note in the context of England, pertained to anxieties about the mass reach of the radio and the extent to which announcers spoke for all of the American population. By the mid- to late 1920s, the U.S. radio announcer had become a sort of cultural idol, with a vocal influence that many took note of. In a 1929 article in the *Los Angeles Times*, Ralph Power noted that "radio naturally is a tremendous potential power and factor in speech education for it enters the American home directly." He asserted that "even more than classroom instructors, the announcer of today exerts a tremendous amount of influence in the language of the growing child."[91] In his 1930 address to the annual convention of the National Association of Teachers of Speech, Henry Bellows, vice president of CBS, also extolled the influence of radio:

> Radio is doing all the time, seventeen hours a day seven days a week, for millions of people what heretofore the pulpit, the stage, and the lecture platform have done relatively infrequently and for a far smaller number of people--it is providing audible models of speech. The models may or may not be good; the fact remains that, good or bad, they are sure to be imitated. People form their speech on what they hear, not on what they read, and certainly today they are hearing more radio than anything else.[92]

Radio announcers were Pronunciphonic subjects par excellence; their words echoed into the ether for vast numbers to hear.

Like the ads for the Pronunciphone, discussions of radio in the 1920s and 1930s suggested that "there is nothing more pleasing than the cultured, American speech and accent."[93] If the conversational elements of the announcer's speech were regimented, the cultured ones were all the more so. During an examination for CBS, potential announcers were required to read the following:

> Among the other prominent musical directors you will hear are Gustave Haenschen and his orchestra, the Detroit Symphony, under the direction of Ossip Gabrilowitsch, featuring Jascha Heifetz and Fritz Kreisler as guest soloists. Ignace Jan Paderewski will accompany a concert featuring the phenomenal youngster, Jehudi Menuhin, while Ernestine Schumann-Heink will sing the Earl King of Franz Schubert.[94]

The *Christian Science Monitor* needed little proof for its suggestion that the modern radio announcer was required to be "a perambulating encyclopedia or the ancient curator of some athenaeum, for whom the entire subject of belles-lettres has become the sine qua non of the intelligent citizen"—it had drawn the quote directly from another announcers' examination passage.[95] A *New York Times* article claimed that only 10 of 2,500 aspirant announcers passed NBC's examination. Along with the range of symphonic terms and names they were required to pronounce, the applicants tended to stumble over the sentence "the seething sea ceaseth and thus the seething sea sufficeth us." Due to its great difficulty, this last sentence was reportedly dropped from later exams.[96]

The mixture of naturalness and formality that characterized the ideal radio voice carried over to an announcer's body and gestures as well. The displaced Emerson College elocution professor Priscilla Puffer claimed that listeners could hear the effects of gesture in a radio announcer's voice. "Though the radio audience cannot see the broadcaster's gesture," she explained in one Emerson College publication, "it unconsciously 'feels' the vital effect which gesture has upon the voice." Unwilling to completely relinquish her elocutionary past, Puffer offered some ornate language in her defense of gesture. "The soul" she claimed, "has only two languages—only two mediums of expression—the voice and gesture." Conceding the naturalness of the new speech, Puffer argued that "the radio broadcaster needs gesture because it is impossible to get the emotional element into the voice unless there is true physical reaction. In order to get a true responsive voice, we first must have a free responsive body." An accompanying picture depicted an Emerson College student gesturing before a radio microphone beside a caption explaining that the student's "gesture was natural, not posed." The ideal radio announcers would need to strike a similar stance, presumably blending their natural movements with the thoughtful embodiment of the elocutionist.[97]

Such discussions of the extraordinary abilities of radio announcers might have suggested that their speech and gesture were out of reach for the everyday speaker. Quite the contrary, the radio announcer was regularly held up as a standard to which every speaker should aspire. Mildred Holland, a nationally syndicated writer whose column, "Making the Most of Your Personality," taught the benefits of a cultured lifestyle, encouraged her readers to emulate radio speech. Stressing the same cultivated friendliness advocated by speech teachers and radio experts, she argued that "some obscure announcers send through the air voices which are so well modulated, clear and intimate that their most commonplace statements are pleasanter to listen to than the valuable and important statements of great authorities."[98] Similarly, a 1929 speech textbook asked students to master a series of tongue twisters similar to those used in radio auditions, including "Amos Ames, the amiable aërialnaut, aided in an aërial enterprise at the age of eighty-six," and "the sea ceaseth and sufficeth us," a variation on the line that had been permanently removed from NBC's announcer exam.[99]

In her column, Holland cautioned that not every radio speaker provided a good model to emulate. Indeed, in addition to the presumably restrained speaking of Graham McNamee and other "ideal announcers," the radio waves carried a variety of more emotionally laden voices. Popular "crooners" such as Rudy Vallee were often targeted by critics for their intensely emotional singing, which departed from traditional norms of masculinity.[100] Others worried that "the 'ain'ts' and 'don't know nothing nohows'" of *Amos 'n' Andy* and other popular radio programs had negative effects on language use, illustrating concerns about both the status of "the King's English" and the effects of an increasing access to African American culture—if only in racist forms of parody.[101] For Holland and similar commentators, the ideal radio announcer served as both an example of the new public speaking and a protection against the other voices that populated the radio waves.

Even for these self-proclaimed protectors of the English language, however, the paradoxical nature of the ideal radio speech caused problems. How exactly should someone strike a balance between friendliness and formality? What was the right amount of personality? The popular "radio priest" Father Coughlin—broadcasting through Detroit station WJR—was chastised at various times by the Catholic Church for his unrestrained speaking. Still, a *New York Times* writer suggested that Coughlin's success came from his "sincerity and an art of speaking to millions in an intimate, appealing fashion." Suggesting his own blend of formality and friendliness, Coughlin claimed that he first wrote his sermons in "the language of a cleric," and then rewrote them, "toning the phrases down to the language of the man-in-the-street."[102] The CBS production director John Carlisle's model radio speaker proved an equally conflicted example. He chose Mussolini, whose voice, he said, "is that of a master of men." According to Carlisle, Mussolini had "the most optimistic-sounding vocal organ ever broadcast. It seems to breathe cheerfulness." In fact, Carlisle said the he "would be glad to have in an announcing staff a voice so distinctive in character."[103]

A *New York Times* article from 1928 further illustrates the conflicted nature of these discussions of the ideal radio announcer. It claims that a successful broadcast speaker "does not strive to reach the minds of his listeners through their emotions," but rather "appeals to their reasoning power to get them into action." The article goes on to suggest that New York governor Alfred Smith found success because "he broadcast a heart to heart talk with the people in their homes," without recognizing how speaking "heart to heart" might itself be viewed as a particular kind of emotional appeal. Similarly, the article recognizes little of the potential contradiction in its headline: "Brevity and Appeal to Reason Make Radio Talks Magnetic."[104] Perhaps magnetism—owing to its associations with electricity—was too scientific to be tainted by emotional artifice.

In spite of these problems and contradictions, in their highly structured speech the ideal radio announcers seemed to their proponents to master the technology of the new media age. Amid the eeriness of the radio waves, these announcers returned with the same intimate authority again and again. The ideal announcer was expected to be sufficiently friendly and sufficiently formal, capable of satisfying the presumably cultured audiences of symphonic music and the more everyday listeners of popular songs. In these ways, the radio announcer modeled the life with technology to which all twentieth-century speakers were supposed to aspire.

Technological Power and Emotional Identities

Despite the insistence of its advocates, there was nothing especially natural about the new public speaking. Experimental speech researchers and radio voice committees alike established extremely narrow rules about pitch, rate, accent, pronunciation, and emotional expression. In fact, as Priscilla Puffer must have recognized, the new public speaking had much in common with the elocutionary speech its celebrants meant to displace; it had replaced the elaborate gestures and phrases of elocution with a new set of carefully defined norms. These conventions were disseminated through public speaking textbooks, research papers, the classroom, newspaper articles, and the speech of radio announcers themselves. Although radio listeners would have heard a variety of voices, and other speaking styles were certainly practiced in public, the continued celebration of this narrow model by speech authorities gave it an important cultural currency.

Part of the power of this model of speech resulted from the close relationships between speech scholars and radio broadcasters, equipment manufacturers, and other professionals who produced a range of products for the general public. In addition to those connections that helped in the manufacture of Seashore's music tests and the Stinchfield-Blanton speech tests, speech scholars had helped the radio industry to create their model of the ideal radio speaker. When WJZ formed its Radio Voice Technique Committee in the 1920s, the New York University public speaking instructors Alvin Busse and Richard Borden used a specially designed "radio recording device" to help the committee establish a model set of vocal qualities.[105] Busse and Borden were the authors of the books *Speech Correction*, *How to Win an Argument*, and *The New Public Speaking*.[106] In *Speech Correction*, Busse and Borden discussed many of the technical and physiological issues that had occupied Glenn Merry and Robert West. Intended for "the mother, the school teacher, the college instructor of public speaking, the family physician, and above all, the *speech specialist*," the book included diagrams of "the pharyngeal cavity" and other organs of the mouth; it also explained how phonograph recordings could be used to correct various speech problems. *The New Public Speaking* and *How to Win an Argument* were handbooks on practical speech for business and other everyday affairs. Busse and Borden were working at the nexus of technological speech research and practical speech production, and building this into WJZ's evaluation of announcers.

Just as the CBS vice president Henry Bellows had addressed a convention of speech scholars, so the academy was turning their attention to researching about and training radio announcers. If Busse and Borden had found a way into broadcasting through their own technological speech research, then it should be of little surprise that Wisconsin's technologically centered speech department would take an interest in radio as well. The Wisconsin speech professor Henry Ewbank chaired the university's radio committee and became actively involved in the university radio station. He also started a course in radio speaking. In an address published in the *Quarterly Journal of Speech* in 1932, Ewbank argued that although many people had intuitive understandings of radio speaking, none of these assumptions could "be finally accepted from the scientific point of view until it has been tested and retested under the best experimental conditions available." The researchers at Wisconsin were attempting to provide this experimental evidence.[107]

Ohio State University was also involved in extensive research and teaching about radio speech. Its School of the Air, sponsored by the Payne Fund, taught a range of courses over the radio, and related research in the speech department and elsewhere sought to perfect this instruction. For instance, the Ohio State professor Frederick Hillis Lumley undertook research on rates of speaking over the radio, as well as the effectiveness of the radio in teaching foreign languages.[108] In 1930, the university also began holding an annual Institute for Education by Radio. The institute featured presentations on the selection and training of announcers, among other topics, and drew university faculty and radio professionals from throughout the country. At the third institute, Wisconsin's Ewbank gave a talk entitled "Methods of Presentation and Speech Suitable for Radio Use." That same year, the famed CBS radio announcer H. V. Kaltenborn gave the opening keynote address, suggesting the practical applications of the institute's various topics.[109]

Similar research and teaching about radio was taking place in other speech departments. Emerson College, having already abandoned elocution for the new public speaking, began offering a course in "radio address" in 1932. The class was taught by Arthur Edes, program director for station WEEI–Boston and a new professor at Emerson.[110] The University of Michigan—where Sherman Lawton was a professor when he undertook his radio speech research—offered its own radio coursework beginning in 1934. The university's department of speech taught classes in "pronunciation, diction and speech delivery under the conditions imposed by the microphone."[111] The University of Southern California's speech department offered courses not only in radio speech, but in speech for "the talking pictures." By 1933, courses in radio were being offered at such institutions as New York University, University of Akron, Pasadena Junior College, University of Iowa, and Kansas State College of Agriculture and Applied Science.[112]

The diverse groups of academics and media professionals involved in constructing the ideal radio announcer were intent on making their mark on the new recording and broadcast technologies. Suggesting the mixture of hope and horror that surrounded these new media, John Peters explains that during the 1920s "many were fascinated and alarmed by radio's apparent intimacy, its penetration of private spaces, and its ability to stage dialogues and personal relationships with listeners."[113] While some simply ranted about the "communized emotions" that flowed from radio and recording technologies, others tried to take advantage of or otherwise control their presumably sublime powers. The narrow vision of human speech offered by the voice experts of speech departments and radio networks provided one means of doing so. For speech scholars, the new public speaking placed their discipline firmly within the modern technologies and sentiments of the twentieth century and divorced them from their Victorian-era roots. Radio broadcasters gave their stations a predictable voice of friendly formality that they believed could serve up orchestral music and advertising slogans with equal facility.

In creating these new models of public speaking, these speech specialists also created a fairly narrow picture of the ideal citizen of the new technological era—one overwhelmingly male, middle-class, and white. Like Elsie McLuhan and Priscilla Puffer, a large number of nineteenth- and early twentieth-century elocutionists had been women, and the speech discipline's rejection of this tradition was in part an attempt to redefine itself as a more masculine enterprise. Criticisms of the hyperemotional nature of elocution were often not so thinly veiled attacks on female speech. According to one commentary written in the journal *Education* when the popularity of elocution was just beginning to wane, "that [the woman speaker] does not pronounce her words correctly, even after the advantages of college education, is true in nearly all cases."[114] The writer advocated the sort of attention to pitch modulation that became important in the Iowa and Wisconsin speech departments. Making the link between speech and emotion explicit, the author ultimately suggested that "there is a moral aspect to the case, as well-modulated voices are often the result of well-controlled emotions."[115] If women speakers lacked appropriate modulations, this author argued, it may have been because of their inability to control their emotions.

The masculine nature of the new public speech was reinforced by the explicit celebrations of the masculinity of the ideal radio announcer. Despite the ambiguity about the ideal announcer's appropriate amount of "personality"—which moved somewhere between McNamee and Mussolini—female speakers inevitably had either too much or too little. The NBC announcer Milton Cross talked about an overly sentimental female announcer who, he said, "seemed to consider that she was intended as a sort of soothing influence to sentimentalize the people." According to Cross, this woman did not recognize that "announcing is just a straight, common-sense, practical job."[116] Bertha Brainard, an early female announcer and station manager for WJZ–New York, suggested other problems with women's speech that she believed were amplified by radio broadcasting. She imagined that one day it would be discovered that "much unhappiness in the world" had been caused in homes and business offices by "women, capable and expert in other ways, [who] lose control and raise their voices in anger or irritation." "In the strident city," she claimed, women "instinctively pitch our tone against the noise all around us. We imagine that if we scream loud enough we can be heard above the racket."[117]

If female voices were too shrill for the radio, however, they were also considered by many to be too boring. According to Phillips Carlin, eastern program manager for NBC, a problem with female announcers was that "women do not speak with authority or conviction."[118] When women *were* able to control their highly emotional voices, offered another commentary, it resulted "in the opposite vice of monotonous, colorless delivery, that of a dead man speaking a dead language." According to this article, "only male announcers, and only a few of them" could "strike the right key, equally remote from Hamlet's ghost and the sweetness of a nightclub hostess."[119]

Of course, these criticisms did not prevent women announcers from taking to the air. Bertha Brainard had apparently found a way to overcome the problems that she herself attributed to female speech. Likewise, another writer observed that despite the comments by various American writers about the inherent weaknesses of the female voice, when Mussolini delivered a special New Year's Day broadcast that was heard in the United States, the "clear, musical, and well-modulated woman's voice" of Signora Boncompagna introduced the orchestral numbers that followed the speech. "Then the question arose, if Italy can accord to a woman the distinction of announcing an international program, then what is wrong with the United States and the chain stations in this country?"[120] Another news article from 1925 noted that women announcers were becoming especially popular in Tokyo, further challenging the supposed weaknesses of women's voices.[121] Katherine Ward Fisher posed a simple question in a letter responding to a *New York Times* editorial about depravities of speech on the radio. Noting that the same issue had told a story of an oratory contest for which the first and second place winners were young women, Fisher asked, "Why aren't there more woman radio announcers?" If women could exceed men on the speaking platform, then why not before the microphone as well?[122]

The voice experts of the period had a range of responses to these challenges to the superiority of male announcers. One commentator described Brainard as the exception that proved the inferiority of female announcers, "the rule rather than a theory." According to him, the majority of women "are rarely found where a knowledge of mechanical devices is essential to the proper achievement of their assigned tasks." "For the most part," he continued, women "follow men to develop projects to the point at which the refining feminine touch is required to give a new industry final polish and luster."[123] After reading Katherine Ward Fisher's comments about female radio announcers, R. P. Jutson, the chief engineer for Radio Centre Inc., wrote his own letter of response. "May I recommend," wrote Jutson, "that the young lady permit herself to study the subject of voice frequency and its action on electrically operated sound reproducers?"[124] As a general rule, it seemed, women were to be polishing technologies rather than working with or speaking through them.

Such commentaries assumed that there was something inherently masculine about recording and broadcast technologies themselves. As Ruth Oldenziel has argued, Americans of the late nineteenth and early twentieth centuries increasingly saw technology as an especially masculine province. During the nineteenth century, American culture had generally placed technology under the framework of "the useful arts," which included a whole range of industrial and manufacturing technologies as well as domestic technologies employed in the home. Around the turn of the century, and with the rise of the profession of engineering, the concept of technology began to replace the useful arts in both professional and popular parlance. As this happened, those innovations designed by and for women were largely marginalized outside the now masculine sphere of technology proper, and women were told with added consistency that the world of the machine was simply not their domain. Technologies were things that men built and worked with, that could presumably not be fully comprehended by women.[125]

This early twentieth-century masculinization of technology fit within a range of other gender anxieties of the period. The concerns about racial and class identity brought about by immigration and urbanization and taken advantage of by stereoscope manufacturers such as Keystone View Company and Underwood and Underwood were accompanied by like worries about masculinity. Anthony Rotundo, Gail Bederman, and Michael Kimmel have demonstrated that the early twentieth-century United States experienced a sense of crisis as the traditional masculinity of the nineteenth century felt challenged in a public that included growing numbers of immigrants, women, and African Americans. While the traditional white, genteel Victorian male could largely assume his dominance based on a particular birthright, the men of the twentieth century would need to prove themselves in a frenetic business and social environment.[126] As Susan Douglas has explained, "for a growing subgroup of American middle-class boys, these tensions were resolved in mechanical and electrical tinkering."[127] Working on various technological projects provided a way of combining more genteel notions of education with working-class ideals of using one's hands. Tinkering was a truly middling practice that brought together science and the shop room into an idealized form of masculine activity.

The emergence of wireless telegraphy provided do-it-yourselfers with excellent opportunities for demonstrating their technological proficiency, and journalists and other writers were quick to champion the masculinity of radio tinkering, especially by young boys. While it was possible to buy completed wireless apparatuses, these writers suggested that the greatest joy and personal benefit came from building one's own. According to the trade magazine *Wireless Age*, building a wireless and then hearing a signal for the first time gave the same pleasure as "the boy's first rifle."[128] Newspaper tales of "heroic boy inventors" such as Walter J. Willenborg provided role models for other boy tinkerers.[129] Orrin E. Dunlap Jr., the radio editor for the *New York Times*, edited a column on radio for the Boy Scouts publication *Boy's Life*, where he made explicit links between radio building and the pursuit of masculine ideals. The apocryphal story of young David Sarnoff's relentless service as a wireless telegraph operator in the wake of the *Titanic* crash no doubt helped to lift him to his future success at RCA. In harnessing the sublime powers of radio, these young men seemingly demonstrated their control over the new media and cultural environment and with it their prowess at a newly technological masculinity.

In attending so closely to radio speech, as well as in their own use of media technologies in the laboratory, the speech researchers of the early twentieth century promoted a similar model of masculine identity. In order to work in Seashore's lab, one needed to be a tinkerer. As I will discuss in the next chapter, this "hands-on" approach to science allowed researchers to distance themselves from the more theoretical and philosophical traditions they were often escaping—a movement itself fueled in large part by concerns about emotion. According to Alison Jaggar, nineteenth-century scientific positivism had forced clearer separations between reason and emotion than had existed at earlier times in Western history. As she explains, "because values and emotions had been defined as variable and idiosyncratic, positivism stipulated that trustworthy knowledge could be established only by methods that neutralized the values and emotions of individual scientists."[130] In Jaggar's analysis, the scientific community created a kind of "emotional hegemony" that championed the very narrow model of emotion encompassed in a certain male, middle-class ideal. The technological approaches of the early twentieth century pushed this ideal still further; in the view of their users, lab technologies both guaranteed the "intersubjective verification" fundamental to a truly scientific knowledge and provided a technological proving ground for a new model of masculine self-control.

While a number of women worked in Seashore's lab and on the technological research at the University of Wisconsin, this did not prevent the sort of emotional hegemony Jaggar discusses. As I have already suggested, it was no mere coincidence that the speech discipline's rejection of the highly emotional elocution paralleled its adoption of a presumably scientific, technology-centered model of speech and research. The new model of "practical" speaking assumed—often quite explicitly, as in the case of Busse and Borden's *How to Win an Argument*—that the proper domains of speech were the brutally competitive business and broadcasting worlds rather than the genteel realm of the speaker's platform. Of course, there was no reason women could not perform well in each, just as there was nothing preventing women from being success-ful speech researchers. However, cultural messages about the mascu-line nature of technology and discussions of the superiority of the male voice in both popular and academic discussions of radio speech no doubt made this success more difficult. Those women who did become speech researchers or radio announcers had to overcome a wide range of stereotypes about their hyperemotional, nontechnological natures.

The presumably masculine qualities of broadcast and research tech-nologies, like the masculinity of the new public speaking itself, pre-sented a still more narrowly defined white and middle-class version of manhood. Discussions of pronunciation and accent were not-so-subtle references to race, class, region, and education. In the Blanton-Stinch-field test manual, the template for evaluating adult speech was based on responses of Italian and German immigrants. The "superior" response for the Italian example, offered after seeing an American bald eagle and flag, had the respondent speaking clearly about his or her memories of posters handed out on President Wilson's arrival in Italy. In contrast, the speaker in the poor response is quoted as saying, "The eagle, just why-er- er- er- just why it was chosen to symbolize America, I don't know. I imagine- er- er- because he is a good fighter." In response to Carl Her-tel's "Jung Deutschland," an image of children in a geography class, the poorest of the German respondents apparently answered, "Th- this pic-ture of the -of the school room, makes me think of the school rooms i- i- in the old country. Th- th- desks are the- are the same,- and- and-and the teacher seems about the same h- h- helpful man that used to t-t-teach me." Stinchfield and the Blantons were making it clear what people they believed were in most need of their speech measurements.

Likewise, just like the stereoscope, the recording technologies of the early twentieth century were celebrated by many for their ability to capture the emotional power of foreign and especially so-called "primi-tive" sounds. A 1916 advertisement for Columbia Records entitled "The Haunting Charm of Hawaiian Music" illustrates a commercial version of this claim. Selling recordings of "the strange sobbing plaintiveness of the voices, the all-but-human notes of the Hawaiian guitar, and the rhythmic throbbing of the *ukulele*," this ad invites consumers to "feel the weird enchantment of life in the south sea islands." The ad contin-ues with an explicitly technological claim:

> The perfect reproduction of Hawaiian music, with all of its strange fascination, is proof of the power and *truth* of Columbia recordings. Test this in *any* form of music—Columbia Records will *prove* it. There is a Columbia dealer near you—let him produce the proof *today*.[131]

Similar to the stereoscopes' international tours, this advertisement used the exotic nature of Hawaii as a means of demonstrating the technological power of the phonograph. That such exotic locales could be held within the grooves of a record or the three dimensionality of a stereoscopic photograph served as proof of the fidelity and tangibility of each.

Such technological and colonialist rhetorics crept into speech and music researchers' discussions of recording technology as well. In an essay entitled "The Collecting of Folk Songs by Phonophotography," Carl Seashore's colleague Milton Metfessel echoed Columbia Records' assertions about the fidelity of the phonograph in capturing "primitive" sounds. "That the ear is inadequate to describe many of the important elements of music," wrote Metfessel, "is best indicated by the American Negro vocal embellishments, whose description has baffled the keenest ear."[132] Owing to the unconventional emotional and musical expressions of African American singing, Metfessel claimed, the "ear analysis" undertaken by most anthropologists had missed important elements of the music. In contrast, using the combination of phonographic and photographic technology developed in Seashore's lab, Metfessel created sound wave photographs of the speaking and singing voices of a group of African American men. Having done so, Metfessel asserted that "the personal decorations of primitive man are no more tangible than the ornaments of voice, when the latter are brought out by phonophotography."[133] Tangibility, here as in the case of stereoscopic tours of the world, was best conveyed through the capturing of what Metfessel and others saw as technologically and emotionally primitive cultures.

Similar technological studies of "primitive" music and speech were undertaken by a variety of other researchers. As early as 1890, J. Walter Fewkes used phonographic recordings to analyze the speech of "the Passamaquoddy Indians," whom he described as "the purest blood Indians now living in the confines of New England." As did Metfessel, Fewkes believed that the "inflections, gutturals, accents, and sounds in aboriginal dialects" could not be captured by conventional means of scientific notation. Having completed his study, however, Fewkes asserted that "the use of the phonograph among the Passamaquoddies has convinced me that the main characteristics of their language can be recorded and permanently preserved, either for study or demonstration, with this instrument."[134] Analyzing phonograph recordings of Chippewa and other native groups, Frances Densmore claimed to have discovered an ancient music all but lost to modern people: silence. "The human race today is forgetting what silence is or can be," she wrote. "The silent figures sitting motionless along the Ganges are monuments to the silence that died centuries ago."[135] Comparable phonographic studies of Native American speech and song took place across the broader social scientific community.[136] This included Iowa, where Seashore and Metfessel were supervising "a program for photographing, recording and interpreting primitive music and speech," for which one student analyzed a recording of "Tlingit Indian speech" provided by Franz Boas.[137]

Like Columbia's commercially available recordings of Hawaiian music, these studies assumed both the exotic, primitive nature of the cultures they captured and the high-tech sophistication of their own recording apparatuses. Although these studies were often cast as noble attempts to hold onto a disappearing culture, just as often they took the perspective of a presumably modern, sophisticated culture looking down on a low-tech, unsophisticated one. Illustrating this outlook, C. M. Wise, having already begun his studies of chest resonance using radio technology, published a discussion of the "Negro dialect" in the *Quarterly Journal of Speech*. Sounding the warnings of the Pronunciphone, Wise ended his essay with the assertion that "the average southern Negro is entirely unconscious of his variant speech, and does not know that improving it would improve his social standing-up to a point where his pigmentation would effectually block further advancement."[138] In order to find social success in the new world of phonographically enhanced language skills, every speaker would apparently need to emulate the middle-class, white voice of Graham McNamee.

In *Bodies That Matter: On the Discursive Limits of "Sex,"* Judith Butler discusses how ideas about rationality and emotion have helped to establish a series of dominant Western cultural conceptions of gender and race. In creating an image of a modern, civilized, rational manhood—of the sort that would be embodied in both the ideal radio announcer and the ideal speech researcher—the dominant voices of Western culture projected upon women, indigenous peoples, and other nonwhites those character traits that they hoped to keep in check. In Butler's words, the traditional masculine "body of reason" requires that "women and slaves, children and animals be the body, perform the bodily functions that it will not perform."[139] In the early twentieth-century climate of American cool, this included especially the bodily and socially dangerous realm of the emotions.

For speech experts of this period, the supposedly hyperemotional, primitive, exotic speech of indigenous peoples and women served as proof of the high-tech, sophisticated, scientific nature of the new, technologically supported speech. On the one hand, women's assumed inability to master the technology of the new media era—illustrated, for instance, in the proclaimed inappropriateness of their voices on the radio—suggested the unique power and rationality of the new, high-tech masculinity. Likewise, the assumedly hyperemotional, primitive nature of indigenous and African American music and speech—like the weird enchantment of Hawaiian music—highlighted the rationality and modernity of speech and music researchers. In capturing the primitive emotions of these voices, these researchers argued, they were doing something impossible before the invention of their research technologies. Thus, in the same way that stereoscopic images of supposedly technologically backwards Inuit groups had demonstrated the sophistication of stereoscope users, the captured voices of indigenous groups and African Americans italicized both the technological prowess of phonograph-wielding scientists and the technological and linguistic backwardness of those they recorded.

Speech scholars, radio programmers, and other self-proclaimed voice experts of the early twentieth century did not necessarily consciously work to recuperate white masculinity or to exclude or "exoticize" women and racial and class minorities. Rather, they were attempting to craft the version of speech they thought would be most beneficial to their students and themselves, as they all made their way through a high-tech new media environment. The new speaker was to dominate this new world through a cool demeanor, a technological proficiency, and a personal but professional voice—all characteristics that had recently been coded as masculine and middle-class. In holding up Graham McNamee as a model for every speaker, the voice experts of this period suggested the extent to which everyone's voice could now be amplified across the public sphere, even as they claimed that only a few voices could survive the new public exposure.

Speaking Softly and Carrying a Big Microphone

The speech of both the ideal radio announcer and the new public speaking embodied the complex hopes and fears about the new recording and broadcast technologies as well as concerns about emotion encompassed in the movement toward "American cool." The diffusion of radios, phonographs, and similar sound technologies throughout the country contributed to the sense of frenetic energy and unbridled emotions that had created uneasiness among authorities throughout the culture. The sublime, seemingly all-encompassing transmissions of the radio carried not only the ideal voices of radio announcers, but the emotional crooning of Rudy Vallee and the sentimental emotionality of jazz, *Amos 'n' Andy*, and radio soap operas. The same technological power that gave voice experts a pedagogical path into homes across the country was also spreading the slang and more colorful pronunciations of the growing popular culture.

This was also the same technological power that seemed to give speech scholars a special access to the inner workings of the human voice. According to Seashore and his fellow researchers, there was no "meaning, or expression of emotion, or art, or skill" in music or speech "that was not represented physically and mathematically in the sound wave."[140] In employing the very technologies that were stimulating the emotions of the general public, these researchers projected an aura of objectivity and high-tech modernity that distanced them from the more sentimental, Victorian perspectives that had dominated speech education during the era of elocution. Like the student of the Pronunciphone, both speakers and speech researchers were to undertake a narrow set of repetitive practices modeled on middle-class ideals of emotional control and decorum.

The conversational but cultured, friendly but formal style of the new public speaking was a negotiation of these contradictory ideas about technology and emotion. The new mass media was dominated by personalities; through the phonograph and radio, a greater number of singers, comedians, and actors were entering people's homes than ever before. In order to be heard among this cacophony of distinct voices, speakers had to have strong, electrifying personalities of their own that nonetheless addressed an audience like an intimate personal friend. The fact that Seashore, the Blantons, Stinchfield, and Gardner and Skinner had found commercial success with their technological speech research highlighted the ways that the commercial, personal, and scientific blurred before the new speech technologies. To find success as a speaker, each citizen would need to be a radio announcer, a scientist, and a hands-on tinkerer, assuming the middle path of emotional control advocated in speech teaching and research and exemplified by the middle-class, white, male voice of the ideal radio announcer.

4

Projecting Emotions

Motion Pictures, Social Science, and Emotional Self-Control

William Marston, one of the pioneers in the development of the polygraph, explicitly celebrated the scientific, cultural, and commercial possibilities of its emotional measurements. In a 1936 letter to a fellow polygraph innovator, John Larson, Marston wrote, "I am in touch with several different fields here where this type of emotion-measurement can be used—and paid for highly—commercially."

> Am also in touch with some big backers of a gorgeous and complicated idea I have for doing this work in many lines on a big scale—I call it THE TRUTH FOUNDATION. My thought is to emphasize the sociological importance of this work—it is far bigger from this angle than anything else ever attempted in psychology. You see, we are attacking the key to moral evil in the world—deception.[1]

Marston appeared to be serious. In addition to his work on the lie detector, he was the creator of the comic book hero Wonder Woman. With her "lasso of truth," Wonder Woman seemed to embody the same moral code that Marston saw embedded in the technology of the polygraph.[2] Harnessing the powers of electricity, the polygraph promised to uncover the truth of human emotions and, in so doing, attack the roots of immorality.

Though extreme, Marston's enthusiasm resonates with the social scientific attitudes about technology and emotion that came to prominence during the 1920s and 1930s. With researchers bringing the powers of electricity to bear on a variety of subjects, many social scientists—like the population more generally—fell prey to the uncritical rhetorics of techno-logical sublimity and media physicalism. As the previous chapter's discus-sion of the public speaking of the 1920s and 1930s demonstrates, speech researchers, like much of the wider culture, were apprehensive about their own and others' emotional control. In response, and as would happen with a range of other social researchers, the speech discipline took refuge in the very technologies it seemed to blame for much of the age's appar-ent emotional intensity. The same technological power that was stimulat-ing the public's emotion seemed to provide the perfect tool for scientific speech analysis. The new recording technologies were both perpetrators of and solutions to the emotional stimulation of the new media age.

These tensions were especially strong with technologies designed specifically to record and measure emotions themselves, such as Mar-ston's polygraph. At the same time that Marston and Larson were devel-oping the polygraph, similar emotion-gauging technologies were being employed in other areas of social research. A range of psychologists, sociologists, and others used technologies similar to the polygraph in their attempts to understand the emotions of their subjects. However, nowhere in these studies did the complex rhetoric of media physicalism appear so vividly as in research that used these technologies to measure the emotional reactions of media audiences. These studies tested the media age's presumed emotional overstimulation by pitting one sub-lime technology against another.

The motion picture studies of the Payne Fund illustrate this espe-cially well. This set of thirteen studies published in 1933 attempted to detail the effects of movies on viewers, especially children. In *Our Movie Made Children*, a summary of the Payne Fund's scientific studies meant for a more popular audience, Henry James Forman found evidence for "the influence of motion pictures and their impersonations upon the character, conduct and behaviour of vast numbers of our nation and especially upon the more malleable and younger people."[3] Forman, like many of his contemporaries, saw both promise and problems in the powerful technology of the motion picture, "second in importance—if second it is—only to the art of printing."[4] The Payne Fund studies sought to measure what had become a commonplace idea, using quan-titative and qualitative methods to demonstrate the influence of the cin-ematic apparatus over the attitudes and emotions of audiences.

This chapter uses the work of Christian Ruckmick, one of the scientists involved in the Payne Fund studies, to explore how the rhetoric of media physicalism took shape for scholars explicitly engaged in researching the emotional effects of media. Ruckmick was the lead researcher on *The Emotional Responses of Children to the Motion Picture Situation*, a Payne Fund monograph he wrote with Wendell Dysinger.[5] Like Forman, Ruckmick and Dysinger showed special concern for the influence of the talking picture, whose "illusion of reality in the theater is so great that to most of the spectators and auditors the presentations carry with them a deep emotional tone, especially in the case of children."[6] Their study utilized equipment developed in Carl Seashore's psychological lab at the University of Iowa, where Ruckmick was a professor and Dysinger a graduate student. Hooking up subjects to psycho-galvanometers and pneumo-cardiographs—which monitored perspiration, respiration, and heart rate—Ruckmick and Dysinger sought to use their laboratory machinery to monitor audience members' emotions. For Ruckmick, Dysinger, and many of their contemporaries, the technological power of such equipment promised a fitting adversary to the power of cinema. These researchers' faith in the readings of the psycho-galvanometer rested on the same belief in the preeminent power of recording technology that drove anxieties about film and radio.

The Emotional Responses of Children to the Motion Picture Situation, as well as much of Ruckmick's wider work, provides a telling example of the tensions about emotion and technology that characterized the media physicalism of this period. Like the public relations specialists, clergy, and film producers of their time, media researchers staked a claim in twentieth-century understandings of emotionality. While Dysinger and Ruckmick's study ostensibly focused on children, they extrapolated their findings to an understanding of the media experience more generally and to a wider theory of human emotions. Their enthusiasm for the physiologically reductive, bio-technological view of emotion suggested by the psycho-galvanometer reflected the anxieties about emotional control that accompanied the growth of entertainment media. Placing an emphasis on emotion-gauging apparatuses allowed these researchers to not only monitor the emotional effects of mass culture, but to project an image of an emotionally detached, empirically rigorous social science that was itself free from the sensational emotions of the new media age.

The consequences of this approach were magnified by the fact that these researchers explicitly distanced themselves from philosophical inquiry and such practices as introspection, both of which apparently carried too much emotional baggage for the new era of emotional control. In their hands, media physicalism was an *anti-philosophical philosophy* that attempted to push aside a range of ethical questions about both the media and their own research practices. Ultimately, and similarly to the discussions in the previous chapters, this research reduced mediated emotion to a set of technical and largely commercial questions and supported the white, middle-class, and masculine conception of emotion that characterized much of the era's technological ideology.

Technological and Social Contexts of Interwar Research

The emotion-gauging technology used by Ruckmick, Dysinger, and other 1920s and 1930s–era researchers fit into a longer history of technology-centered inquiry. Scientific measurements played an important part in nineteenth-century eugenics. As I mention in chapter 2, nineteenth-century scientists had used photographic technology to analyze a range of physiognomic features thought to demonstrate a person's character. Craniometers, cephalometers, calipers, and other apparatuses had also been used to measure people's presumed racial and sexual characteristics. The measurements of cardiographic pressure that would be drawn from the pneumo-cardiograph used in the Payne Fund studies had taken place at least as early as the 1870s when the French scientists J. B. A. Chauveau and E. J. Marey measured the beats of a horse's heart.[7] The Italian scientist Angelo Mosso used a variety of instruments—a number of which he designed himself—to measure his own emotional changes, reporting that "his own thermometer-measured rectal temperature changed with his spontaneously evoked emotion."[8]

The nineteenth-century German researchers Hermann von Helmholtz and Wilhelm Wundt used similar technologies to study a range of perceptual issues. Helmholtz, a physician and physicist, used telegraphy as a model for the human nervous system. He also created instruments such as the ophthalmoscope and resonator to measure and study sounds, making important strides in the psychology of acoustics.[9] Wundt, who established the first formal psychological laboratory in 1879, extended

Helmholtz's research on sensory perception. In contrast to many of the early twentieth-century U.S. psychologists who took up these experimental techniques, however, Wundt was equally committed to philosophical methods of exploration. As Kurt Danziger has explained, "for all his interest in experimentation Wundt was sufficiently immersed in the German tradition of objectifying *Geist* to reject the notion that a study of the isolated individual mind could exhaust the subject matter of psychology."[10] According to Danziger, Wundt "had no interest in the possible practical applications of psychology and he had no interest in converting psychology into an independent discipline without ties to philosophy."[11] Owing to his commitment to philosophy, Wundt explored a range of questions in the psychology of aesthetics and music, as had Helmholtz before him. In fact, Wundt and Carl Stumpf—whom Christian Ruckmick called "a pioneer in the field of psychology of music"[12]— engaged in a four-year debate about "music consciousness" (*Musikbewusstsein*) aroused by disagreements regarding the relationship between musical aesthetics and the psychophysics of sound.[13]

In their discussion of scientific image making in the late nineteenth and early twentieth centuries, Lorraine Daston and Peter Galison offer a related narrative regarding the scientific struggle over technology and objectivity. Focusing on the creation of scientific atlases—"the bibles of the observation sciences"—Daston and Galison explain the wide range of machinery scientists employed to guarantee the fidelity of their representations of nature.[14] As they explain, scientists turned to such technology as the camera obscura and photograph for reasons of both accuracy and morality; the nineteenth-century scientists Daston and Galison explore saw it as their duty to avoid imposing their "hopes, expectations, generalizations, aesthetics, even ordinary language on the image of nature."[15] Apparently free from human subjectivity and desires, machines promised scientists a truer representation of the natural world. As Daston and Galison note, "while much is and has been made of those distinctive traits—emotional, intellectual, and moral— that distinguish humans from machines, it was a nineteenth century commonplace that machines were paragons of certain human virtues."[16]

Early twentieth-century social science saw the virtues of machines expanded to include the empathetic access to human emotions. If Marey and other scientists had used technology to negotiate their anxieties about objectivity, a number of early twentieth-century researchers used it to navigate the complex scientific and cultural climate of early twentieth-century emotion. As had been the case with speech scholars' criticisms of elocution, a number of self-proclaimed experts on the scientific and medical aspects of emotion argued for the benefits of restraint for a whole range of conditions. If speech scholars saw emotion as hazardous to people's vocal performances, this larger group of psychologists, sociologists, and medical professionals saw them as dangerous for people's bodies more generally. Emotional restraint was important medically, scientifically, and socially.

This climate of restraint provided the backdrop for the technological take on emotions that became especially prominent during the 1920s. At this time, instruments of measurement found widespread use in the social sciences, as Mark Smith's discussion of the Social Science Research Building at the University of Chicago helps to illustrate. Dedicated in 1929, "Eleven Twenty-Six" had only two lecture rooms, and "most of the other rooms were filled with galvanometers, calculators, and the like to enumerate data. There was no space at all for books."[17] Similar technologies had been used by earlier scientists such as Wundt, Marey, and Mosso; however, as Otniel Dror explains, "it was not the novelty of the technologies that made the new science, but their radically new application and the innovative interpretation of their by-now-familiar inscriptions, tracings, and outputs."[18] Further, Dror suggests, "by the 1930s the use of various instruments and laboratory procedures to represent emotions was widespread in Anglo-American culture."[19]

This technological emphasis complemented a series of other social scientific developments taking place in the 1920s and 1930s, which themselves reflected concerns about emotional restraint. An increasing suspicion of "introspection" meant that researchers were less likely to draw on their personal experiences or to have subjects describe their own sensations. Just as scientists of the emotions such as Angelo Mosso had experimented on themselves, researchers in psychology had made heavy use of their own subjective experience, seeking to understand psychological processes through careful introspective investigation. Drawing on psychology's philosophical heritage, in 1894, Yale's George Trumbull Ladd argued that "much skill and success may be attained by intelligent practice in the analysis of one's own mental states with the instrument of introspection."[20] For Ladd and others, careful thinking about one's own thoughts was considered a legitimate path to psychological knowledge.

As psychologists worked to differentiate their work from that of spiritualists and philosophers, and to position psychology as a legitimate science, introspection gave way to more exclusively objective, technologically focused research.[21] Instead of pursuing the squishier questions of mind and spirit, psychologists increasingly confined themselves to those phenomena observable to the laboratory apparatus. The economist and "Eleven Twenty-Six" resident Wesley Mitchell similarly urged social scientists to draw on new developments in statistics to "advance their knowledge as rapidly as possible from the stage of speculations about hypothetical conditions to the higher stage of quantitative analysis of actual conditions."[22] Psychology was no less taken by this move toward quantification, and introspection was increasingly seen as an impediment to the discipline's ascendance to this higher stage of analysis.

The American uptake of Freudian psychoanalysis contributed to these social scientific developments in some interesting and complicated ways. In *Civilization and Its Discontents*, Freud had himself discussed some of the emotional effects of the era's new technologies.[23] Likewise, his theories of the unconscious suggested a number of ways that people's psychic lives could distort their sense of reality, reinforcing the concerns about emotions that pervaded much of the early twentieth-century social sciences. The American social scientist Harold Lasswell turned Freud's theories into a more thoroughly technological, quantifiable approach. Lasswell had been mentored by Elton Mayo, the psychologist responsible for the Hawthorne Studies. Taking his own technological approach to psychology, in the 1920s and 1930s Mayo had varied a range of physical conditions in the Hawthorne Works factory, including adjusting the lighting in incremental ways. Lasswell worked with Mayo at the Boston Psychopathic Hospital in the 1920s.[24]

Lasswell, who while studying in Austria made a special trip to meet with Freud's daughter Anna,[25] eventually employed emotion-measuring equipment as part of the psychoanalytic interview process itself. One of the problems with psychoanalytic practice, Lasswell argued, was that there was no standard way for psychoanalysts to record the case histories they took from their patients. This applied not only to the recording of a patient's words, but to the patient's physiological changes as well:

The phrase "restless movement of the hands frequent" may mean that the subject, whenever observed by the interviewer, was fumbling with his belt during the first hundred hours; or simply that the interviewer vaguely remembers that the subject occasionally pulled his nose or scratched his head.[26]

According to Lasswell, this sort of vagueness could be overcome by appropriating "the objectifying procedures of general psychology."[27] For Lasswell, recording a patient's physiological responses via technology offered just this solution, supplementing the qualitative information of the interview with the more quantitative, statistical data increasingly being celebrated in American psychology. At the end of a session, psychoanalysts would have both their own notes and a presumably objective record of a patient's emotional responses. From Lasswell's perspective, the emotion-measuring power of these technologies made them more perceptive than psychoanalysts themselves.

The sociological research of the 1920s and 1930s saw a gradual movement toward "scientism" and quantification as well.[28] This version of sociology "increasingly embraced the example of the natural sciences,"[29] aiming to "confine itself to the observable externals of human behavior"[30] rather than some of the more speculative normative and ethical questions with which earlier sociologists had grappled. As did the speech discipline's movement from elocution to public speaking, the trend toward empirical methods and away from introspection and philosophy played into the cultural anxieties reflected in the movement toward "American cool." Sociologists and psychologists aimed to demonstrate their objectivity by showing that they were beyond the emotionally tainted pasts of their own disciplines as well as above the emotional stimulation of the new media age.

A variety of research on media reflected this growing concern with emotion as well. In 1916, Hugo Münsterberg offered one of the earliest psychological studies of film: *The Photoplay: A Psychological Study*. A student of Wilhelm Wundt, Münsterberg was quite comfortable applying psychological methods to the aesthetic experiences of art forms. Beginning his studies in Wundt's laboratory in 1883, Münsterberg earned his degree before the widespread founding of psychological laboratories in the United States. As a result, his education avoided some of the more explicit disciplinary quarrels that took place in the early twentieth-century United States as psychology split away from its parent discipline. Like his teacher, Münsterberg saw value in the linkages of these two areas of study.[31]

Münsterberg's *Photoplay* bears the marks of a scholar more comfortable with both overt emotional stimulation and philosophical inquiry than Ruckmick and many other later psychologists of film. At the opening of one chapter, Münsterberg says explicitly that "to picture emotions must be the central aim of the photoplay."[32] In his view, the new technologies of the cinema allowed for more creative aesthetic techniques that could challenge audience members with a range of emotions they had not experienced in other art forms. Owing to its possibilities for violating time and space, Münsterberg argued that motion picture technology could far surpass the theater in stimulating emotion. For instance, a shaky camera could produce "a feeling of dizziness" and "an uncanny, ghastly unnaturalness."[33] Despite his hope for the cinema, however, Münsterberg doubted that his contemporary filmmakers would provide viewers with these challenging aesthetic experiences. Because the motion picture had "not yet emancipated itself sufficiently from the model of the stage,"[34] Münsterberg lamented that this emotionally powerful cinema "still belongs entirely to the future."[35]

In addition to these more strictly psychological investigations, Münsterberg posed a series of questions about aesthetics firmly rooted in his philosophical background. In "The Purpose of Art," a chapter following his earlier discussions of emotion, Münsterberg pulls together examples from Beethoven, *Antigone*, *Hamlet*, and Rembrandt. In "The Means of the Various Arts," he draws similarities and contrasts among a range of traditional arts, including painting, novel writing, sculpture, poetry, theater, still photography, and musical composition. Throughout, Münsterberg makes clear that he does not intend simply to analyze the effects of the movies. His book offers an argument about the aesthetic power of cinema, attempting to defend it from those of his contemporaries who saw it merely as a form of entertainment with no artistic value. Both his more psychological and philosophical discussions celebrated the unique abilities of motion picture technology to stimulate human emotions.

By the 1920s—at the same time that social scientists were employing progressively more recording technologies in their labs—academics increasingly seemed to take aim at the emotional power of movies.

The criminologist John Oliver worried that "under the strain of" an emotional performance, audiences were "no longer able to make use of [their] thinking centers." He suggested that people test his hypothesis by telling an audience member mesmerized by an emotional scene that "a man behind you has just dropped dead," or by handing them "some chocolate-coated lumps of cotton" to chew on, noting that they would likely "chew away" until the "emotional stress lessens." Imagining a hypothetical situation with a female viewer, he continued that "if now, in her emotional state, some impetus rises in her to do some act that she would ordinarily shrink from, she will surely follow that impetus as it rises from the lower centers of her being, follow it mechanically, blindly, without consciousness of what she is doing." Oliver saw this effect as a direct impact of emotion on the body: "In emotional stress, whatever the immediate mechanism may be, the result is that the lower, more mechanical, animal centers in the medulla and spinal cord dominate and overpower the higher centers of the brain."[36] Sitting in a theater apparently turned spectators into emotionally driven animals.

Joseph Geiger, a professor at the College of William and Mary, expressed similar concerns, worrying about the same technological sophistication that Münsterberg had praised. "There is a vast psychological difference between hearing or reading an account of a murder, or an assault, or a passionate mutual attraction between members of the opposite sexes, and seeing these things actually portrayed on the screen,"[37] Geiger wrote in 1923. As a result of their visual and technological power, he further explained, movies "induce an overstimulation of [children's] imaginations, resulting in a condition of nervous excitement which cannot but have an unfavorable effect on their general health."[38] Harmon Stephens of the University of Tennessee argued that the film industry's "progress in both artistry and technique has been marvelous." However, he added, "improved artistry and technique have often provided attractiveness for questionable things which crudeness formerly left uninviting."[39]

Münsterberg, Oliver, Geiger, and Stephens shared a belief in the emotional power of film technology, even if they evaluated that power differently. Both positions reflected longer-standing concerns about photographic images. Nineteenth-century still photography had been shrouded in its own veil of mystery. The disembodied images of early photographs—made all the more eerie by "ghost images" that resulted from camera movements and other interruptions in the photochemical process—encouraged popular and scientific interest in "spirit photography" and other psychic practices.[40] A similar sense of strangeness surrounded motion picture technology. During a movie screening in Canonsburg, Pennsylvania, in 1911, the film apparently snapped and sent a bright white light across the screen. A riot began when a frightened group of boys yelled fire and started a stampede among the fleeing audience members, leaving twenty-six dead and a number of others injured.[41] This incident seemed to suggest that not only the content, but the mysterious technology of the motion picture itself could evoke dangerous emotions within the public.

Such incidents reflected concerns about the sublime powers of cinema shared by Geiger, Oliver, Stephens, and a growing number of others. In a 1921 essay in the *North American Review* entitled "The Movies as Dope," Elizabeth Robins Pennell agreed with Oliver's concerns about the mesmerizing effects of motion pictures. "The evil they work," she wrote, lay "in the state of Nirvana to which they seduce their audience."[42] A *New York Times* article in which one man lamented that "we expend so many maudlin tears over the absurd sentimentality of the movies" also included the thoughts of the psychiatrist A. A. Brill. According to Brill, "those races which have learned to control their feelings are undoubtedly of a higher type of civilization than those which have not."[43] Widespread calls for film censorship demonstrated that many of the era's educators felt that the movie industry was betraying this higher civilization of emotional and moral control.[44] Despite these growing criticisms, however, a number of people continued to celebrate the cinema's emotional power. In a 1931 article in the *New York Times*, the Philadelphia conductor Leopold Stokowski speculated that "over sound films of the future I believe we will be able to convey emotions higher than even thought—things subtle and intangible—almost psychic in their being."[45] The same psychic powers that excited Stokowski were precisely what worried a number of film's critics.

These ambivalent ideas about technology were important fodder for the movies themselves. Fritz Lang's 1927 film *Metropolis*, which was released in an edited form in the United States, offered a strong critique of the impact of technology on emotion. The film begins with images of spinning gears and clock faces against the backdrop of a futuristic city. Suggesting the numbing effects of the city's technologies, lines of uniformed workers—their heads bowed—march in lockstep into the bowels of the hall of machines. At the same time, and in line with much of the criticisms listed above, the film suggests that technology can have an overstimulating effect on the emotions. The wealthy classes of the city live in a decadent "club of the sons," filled with dancing girls and other theatrical frenzies. Both the wealthy and working classes of the city are ultimately seduced by a robot—the silver-bodied machine that is eventually given the form of a beautiful woman by its inventor. Watching the film, Herman Scheffauer wrote in a *New York Times* review, "we feel that new forces have been unloosed upon this earth of ours; that man is being enslaved by them, even the man who is master of the human and mechanical slaves."[46] Of course, the film's own "miraculous" photographic effects and status as an unparalleled "technical marvel"[47] relied upon the very technological sophistication it was presumably critiquing.

Released in 1933, the year the Payne Fund studies were published, the high-tech Hollywood production *King Kong* offered its own commentary on the modernizing effects of technology. As the film opens, Carl Denham, a film director known for spectacular pictures, takes a crew onto the remote Skull Island in search of the legendary Kong. When they find him, Kong is being treated as a god by the villagers of the island, who offer him human sacrifices through the door in the giant wall enclosing his island refuge. When Denham arrives, Kong absconds with Ann Darrow (Fay Wray) and then fights a series of giant, prehistoric monsters, before Denham and his crew capture him and return with him to New York City. Displaying Kong live and in chains for a New York audience, Denham explains that "he was a king and a God in the world he knew, but now he comes to civilization, merely a captive, a show to gratify your curiosity."

Suggesting the uncomfortable relationship between this "beast" of nature and Denham's civilized, new media world, Kong is so angered by the repetitive camera flashes of the news reporters who try to take his picture that he breaks free from his chains. This sets off Kong's famously destructive rampage through New York City, in which he smashes subway tracks, tosses cars, and climbs skyscrapers, whose windows frame his frightening face in testament to the tensions between wild nature and civilized technology. In the end, Denham tells reporters that Kong was killed not by the airplanes that shot him down, but by his debilitating love for Ann Darrow, whom he takes with him to the top of the

Empire State Building. Yet the movie ultimately suggests that both might be true. On the one hand, Kong's destruction results from Denham's own obsession with mastering nature through technology, as he uses his movie camera and gas bombs to capture Kong for the edification of audiences seeking increasingly exciting entertainment. But, as Denham himself suggests, Kong also falls as a result of succumbing to his emotional attachment to Ann Darrow. In the end, Kong's death offered support for both the sometimes dangerous superiority of newer technologies and the necessity of emotional control.

Similar concerns about emotion and technology played an important role in the Payne Fund studies, as well as in the work of Christian Ruckmick more generally, who, as far as motion pictures were concerned, sided overwhelmingly with the pessimism of Geiger, Oliver, and Stephens, and against the optimism of Münsterberg. The Payne Fund researchers' interest in emotion was a product of both developing cultural and social scientific attitudes about self-control and the larger context of the Payne Fund and early twentieth-century social research itself. The Payne Fund had been founded by Frances Payne Bolton, a wealthy philanthropist who, like the Carnegies and Rockefellers of the time (who were also supporting social research), had a special concern for matters of public health. Before creating the Payne Fund in 1927, Payne Bolton had helped organize the National Committee for the Study of Juvenile Reading in 1925. This group focused on the emotional impact of another popular media, exploring "the comparatively recent wave of magazines, reaching millions of copies a month, devoted to sensationalism in depicting temptations, narrow escapes, and regrets."[48]

This public health approach inflected the Payne Fund's projects and outlook as a whole, reiterating some of the notions about emotional addiction and mania that accompanied such technologies as the automobile. Ella Phillips Crandall, the secretary of the Payne Fund and a member of the fund's national committee, was the former director of the American Child Health Association; the ACHA's director of publicity and promotion was also on the Payne Fund's national committee. Frances Payne Bolton herself was invited to join the Association of Women in Public Health.[49] Seeing health issues in a global context, in addition to the fund's projects on reading and motion pictures, the group also supported research on opium use in Europe. Each of these topics represented its own sort of cultural and bodily pollution. As Payne Bolton explained in a discussion of the fund's media research,

> Since the war, when I was thrown into close contact with certain groups of youngsters, I had a realization of what a tremendous influence reading matter and movies have—I have been horrified at the type of material available. It would seem as if all the evil influences of the world were concentrating on the ruining of the morale of youth—and I started in to see whether the facts could be gotten together and something done.[50]

In the view of the fund, sensational media and opium were not so far removed from each other.

The final thirteen Payne Fund motion picture studies primarily targeted the overly stimulating crime, adventure, and romance films believed to be especially corrosive to the American public. In his Payne Fund summary, Henry James Forman recognized that crime was a fact of life. And yet, he added,

> That it is emphasized on the screen out of all proportion to its place in the national life is equally clear of doubt, indeed, glaringly obvious. Were crime to receive similar emphasis in the life of any one of us as individuals, we should properly expect to be either in jail or in an insane asylum.[51]

Edgar Dale, a research associate at Ohio State University's Bureau of Educational Research and the author of three Payne Fund monographs, drew explicit parallels between romance and crime films and drugs, reiterating the larger attitude of the Payne Fund as a whole. Both drugs and movies were a form of escape and thus served a social purpose. However, if drugs should be used only "when nothing can be done for the patient except to relieve him of pain," then why should dangerously themed movies provide children's escape when other forms of recreation were available? A "well-balanced motion picture diet" would have far less sex, crime, love, and mystery films than the typical Hollywood movie season.[52]

Despite this general apprehension, however, the Payne Fund suggested some positive possibilities for the developing motion pictures. Frances Payne Bolton and the other fund administrators were in regular contact with George Skinner, who was developing a project on visual education that would provide films to schools. Although the fund would not ultimately partner with Skinner, the Payne Fund studies suggested that motion pictures could have an important place in education. The visual power of films made them ideal for teaching information to children, because the compelling imagery of the picture made the communicated information easier to retain. Maintaining their basic ambivalence about the movies, however, the Payne Fund researchers ultimately argued that most motion pictures were providing *disinformation*, and thus distorting these educational possibilities.[53]

The Payne Fund had a similarly ambivalent approach to radio. Frances Payne Bolton told RCA's Franklin Dunham that she "would hate to see the air polluted as the motion picture had been," but said that she "felt it was well on the way."[54] But the fund was also making heavy use of radio as an educational tool. As I mention in the previous chapter, the Payne Fund supported research on radio education at Ohio State University and was a chief backer of the Ohio School of the Air, started by Benjamin Darrow in 1928. Darrow had begun in radio while working as the director of the Boys and Girls Division of Sears Roebuck and Company's Agricultural Division. He was "Uncle Ben" for a radio program broadcast on Ohio's WLS. After making his fortune on the TABL-TUB, a table that converted into a bath or washtub that he patented as a product for agricultural families, Darrow decided to devote his time to educational radio. He teamed with Armstrong Perry, a radio writer and Boy Scout leader who made regular contributions to *Boy's Life*. Darrow and Perry planned to provide radio broadcasts for use by teachers in schools as well as educational programs for people without access to classrooms.[55]

Darrow himself offered conflicted messages about the emotional power of radio technology. He wrote that "children are listening to the radio so much that a person is loathe to believe the figures until they are proved." Of course, his school radio program would result in children listening to more, not less radio. Like the Payne Fund that supported him, Darrow worried that the current radio programming hurt children's emotions by delivering "a too-heavy diet of mystery and murder, suspense and horror."[56] This resulted in "what may be an overemphasis even to older children, an over-supply of evil." Even as Darrow expressed concern about radio's corrupting influence, it was precisely the medium's impact on the emotions that seemed to make it a valuable educational tool. In an evaluation of Darrow's School of the Air, J. L. Clifton, director of the Ohio State Board of Education, stressed the influence of radio on the emotions as one of its chief educational benefits. "Many teachers are convinced that children are emotionally affected by the knowledge that they are listening in company with thousands of other children," wrote Clifton in 1930.[57] Christian Ruckmick apparently shared some of this optimism about radio education. In the 1930s, he offered a course entitled Systematic Psychology: Emotion over the University of Iowa's radio station WSUI.[58]

Ruckmick's understanding of the motion picture, which to a large extent he shared with his fellow Payne Fund researchers as well as much of the wider social scientific community, reflected these conflicted celebrations and denunciations of technology. For Ruckmick and many other researchers, the emotional power of recording technologies made them as dangerous to audiences as they were useful for social scientists. Hugo Münsterberg had challenged film artists to use cinematic technology to enhance viewers' emotional stimulation. Ruckmick and Dysinger used the emotion-gauging apparatus of the psycho-galvanometer to demonstrate the importance of emotional restraint to both movie audiences and communication research itself. By mediating a subject's emotions through a constant flow of empirical data, the psycho-galvanometer promised to remove the human scientist from the process of emotional interpretation even as it provided intimate access to a subject's innermost feelings. This perspective built on the technology-centered tradition of the University of Iowa's department of psychology.

Christian Ruckmick and the Iowa School's Psychology of Art

Driven by Carl Seashore's studies of music, Iowa produced an impressive corpus of art-related empirical research from the earlier twentieth century up through and beyond the 1930s. During this time, faculty and graduate students such as Seashore, Ruckmick, Dysinger, Norman Meier, Milton Metfessel, Jacob Kwalwasser, and Grace Helen Kent developed a variety of apparatuses to measure the emotions of both artworks and audience members. These studies were published widely in psychological, educational, and scientific journals, and a series of special issues of the *Psychological Review* were dedicated to *University of*

Iowa Studies in Psychology, which focused predominantly on these art-centered, technologically driven experiments. This research assumed that well-designed equipment could capture the essence of the emotional elements of an artistic experience and thus allow the scientist to faithfully measure and manipulate it.

Seashore had entered Yale's doctoral program in philosophy in 1892, the same year the philosophy departments at Yale and Harvard founded their first psychological laboratories. Unlike Münsterberg, who took his degree when psychological research was a smaller part of the discipline of philosophy, Seashore was educated as psychology was coming into its own as an independent discipline. At Yale, Seashore studied primarily with the philosopher George Ladd and the experimental psychologist Edward Wheeler Scripture. According to Seashore, Scripture "was treated as a technician" by the department's philosophers, even as Scripture himself "conveyed the impression that the systematic statements in the learned books of the day might all be true, but they were nevertheless second-hand" until they had been experimentally tested.[59] Seashore's education was an ambivalent balance of philosophy and the new psychology.

Iowa's psychological laboratory had been founded in the 1890s in the department of philosophy by George T. W. Patrick. Less inclined toward experimental than theoretical research, in 1897 Patrick, an Iowa alum who had also earned his PhD from Yale, hired Seashore to help run the laboratory.[60] Like Münsterberg, Seashore saw philosophical questions about aesthetics as closely intertwined with those of experimental psychology, even if he tended to place his emphasis on experimental investigation. This blend of philosophy and psychology was most clearly illustrated in his experimental studies of the psychology of music.

Seashore began testing students' perceptions of variations in pitch in response to a music colleague who boasted of his superior "musical ear."[61] A resulting study was published in 1899 as part of the *University of Iowa Studies in Psychology*, then edited by Patrick.[62] In the next volume of Iowa's *Studies in Psychology*, Seashore introduced his "voice tonoscope," which, he wrote, could be "used in measuring the pitch of the human voice in singing and speaking."[63] Seashore took over the editorship of the series with volume 4, published in 1905 as part of the *Psychological Review*. In volume 6, published in 1914, all seven of the collection's essays explored music-related topics. It included one study of the sense of rhythm and two studies each on pitch discrimination, variations among musical tones, and apparatuses for measuring and analyzing sound.[64]

A 1916 essay introduced Seashore's "tonoscope"—dropping the "voice" to illustrate that it could be used to measure a wider range of sounds. It offered these measurements through the use of "stroboscopic vision, the principle of moving pictures."[65] The tonoscope transformed sound into visual data: the frequency of a sound vibration caused a flame to raise and lower and to illuminate specified matrices on a rotating aluminum drum. This resulted in a visual graph of a sequence of sounds, showing the peaks and valleys of vocal or instrumental tones. Charting sound in this manner, Seashore argued, allowed scientists a powerful means to objectively investigate the emotions of music. In another essay, Seashore explained that his research sought to "eliminate subjective and circumstantial conditions and accessory features . . . and consider only beauty as it is objective in the physical tone." This objective beauty, he further asserted, "may be measured with considerable precision in terms of the form of the sound wave."[66] By empirically charting sound waves—which carried all the musical and emotional elements of a piece of music—the tonoscope and similar pieces of equipment captured music's beauty and emotional power. "By the process of photographing sound waves," Seashore believed, "we obtain adequate data for an objective portrayal or description of the expression of emotion in music."[67]

By capturing sound waves empirically, the tonoscope and similar instruments allowed Seashore and his fellow researchers to study art and emotion while still holding to their ideal of an objective social science. Recognizing that "experimental psychologists have often avoided aesthetic problems because matters of feeling are intangible," Seashore aimed to bring tangibility to feelings and aesthetics through technological mediation.[68] Seashore understood that such an empirical approach to music "sounds mechanical and oversimplified," but held that it was actually "marvelously beautiful."[69] In fact, he maintained that his technological approach to music worked so well that scientists knew the true beauty of music better than the casual listener and even better than musicians themselves:

The musician waits for the psychologist to blaze the trail. He is a most docile inquirer when opportunity is given. The perspective of music, and the perspective of the musician, which is gained by the objectifying of factors involved, will be projected into our common account of music, and this will vitalize musical ideas and furnish the singer a more general insight into his capacities and possibilities.[70]

Seashore's approach to music was neither a touchy-feely aesthetic of intangible emotions nor a cold, calculated scientific abstraction. As objective as any scientist and as musically sensitive as any musician (even more so), Seashore believed that his equipment gave him an empirical gateway into the deep emotions of music, all of which lay buried in the sound wave.

Seashore's work gained wide recognition. He served on the selection committee for several prestigious awards and earned honorary doctorates from such schools as Yale, the University of Southern California, and the University of Pittsburgh.[71] As a testament to the success of his empirical approach to emotions, Seashore was invited to participate in the 1927 Wittenberg Symposium on Feelings and Emotions at Wittenberg College, for which he presented a talk on the place of phonophotography in the study of emotion. This symposium brought together thirty-three established researchers on emotion, including Walter Cannon and Alfred Adler. Wittenberg College president Rees Edgar Tulloss opened the meeting by observing that "the time is now ripe for a gathering of the forces of psychology for an attack upon this important field," arguing that "the form of attack most likely to achieve results of consequence is that of experimental procedure."[72] Further illustrating his prominence, it was Seashore and his work that initially drew the attention of the Payne Fund to Iowa.[73] The technology-centered approach to emotion undertaken in Seashore's lab was finding strong traction in psychology.

The more audience-centered studies of Ruckmick, Dysinger, and others at Iowa shared a technological orientation and basic set of assumptions with Seashore's work. Ruckmick had followed Seashore's work before joining Iowa's department in the mid-1920s.[74] In 1928, he replaced Seashore as the editor of the *University of Iowa Studies in Psychology* series, where he continued the psychology of art perspective that had blossomed under his predecessor. Prior to his move to Iowa, Ruckmick had already turned his attention to music and the developing area of communication media. His doctoral work at Cornell focused on individuals' abilities to perceive musical rhythms, employing metronomes, strobe lights, and pneumographs to analyze the bodily changes brought about by rhythmical experiences.[75] While teaching at Wellesley College he proposed an Institute for Acoustic Research, recognizing that "we have lately witnessed an immense development in the photographic and cinematographic industries; now comes the prospect of an even wider application of wireless telephony."[76] Curious about the impact of emerging technologies on the well-being of his subjects, Ruckmick developed and utilized a wide range of laboratory apparatuses to study these effects, ultimately seeing emotions themselves as particular kinds of technological impulses.

Like John Oliver, Joseph Geiger, and A. A. Brill, Ruckmick approached emotion as something that needed to be thoroughly managed. Emotions were a kind of bodily "commotion"[77] that could do profound psychological, biological, and cultural damage if allowed to flourish uncontrolled. In an essay entitled "Why We Have Emotions," Ruckmick stressed the importance of a decidedly average, restrained level of emotion. "In all time, not only in our time, emotionless fiends as well as frenzied fiends have claimed their victims. This human flotsam and jetsam soon came into the focus of scientific scrutiny and new questions were raised in the thinking world."[78]

Elsewhere, Ruckmick is similarly explicit that restraint and control are imperative to bodily and psychical health. In a letter to William Short, executive director of the Motion Picture Council and an advisor of the Payne Fund studies, Ruckmick offered that "continuous or frequently repeated high emotional stress leads to a neurotic condition of the human organism."[79] Although Ruckmick recognized that some degree of emotional experience and expression is normal and healthy, he made clear that excess feelings should be avoided:

> We commonly say that certain people crave more excitement. This is not mentally hygienic and it results in a social structure keyed up only on one side. Even a democracy may become frivolous, in which case the end is bound to be bad because the select few who are elected to positions of honor and power could appeal only to this side of human nature.[80]

Emotional restraint was fundamental to both healthy bodies and healthy democracies.

Along with his concerns about the emotional health of his psychological subjects, Ruckmick was also concerned about the emotional control of scientists and other researchers. Ruckmick praised scientists who purged emotional traces from their writing. In a letter to Walter Cannon, Ruckmick celebrated the way Cannon had refuted another scientist's findings: "I want to congratulate you on the impersonal and wholly logical fashion in which you couched your reply."[81] Later, in a review of the book *The Fifth Column Is Here*, by George Britt, Ruckmick noted the book's "flamboyant cover" and the "high powered journalistic style" through which Britt "stirs up a considerable amount of emotional appeal." Commenting that Britt "almost shouts at us throughout his book," Ruckmick concluded that "in some cases we may even doubt whether the author's interpretation of the facts, due to his own alarmed state of mind, is not somewhat overdrawn."[82] Extreme emotional expressions were certain to cloud one's thinking. For Ruckmick, good research required an emotionally controlled demeanor.

This suspicion toward emotion had profound methodological implications for Ruckmick, as it did for other psychologists of his time, particularly in terms of their movement away from the introspective methods that had served the discipline during its earlier stages. If emotions were bodily disturbances (another of Ruckmick's terms and the predominant phrase used in the Payne Fund studies) with a potentially negative impact on a person's judgment, then how could subjects be expected to introspect upon themselves in a meaningful manner? Earlier researchers had taken for granted that introspection was both possible and useful, but saw their perspectives change with the more exclusively empirical emphasis of the 1920s.

Ruckmick's own research is a telling example of this. While he began his work as someone heavily steeped in introspective analysis, he eventually saw it as a thorn in the side of a truly scientific psychology. At Cornell, Ruckmick had studied under Edward Bradford Titchener, one of the champions of introspection. Like Münsterberg, Titchener had been a student of Wilhelm Wundt, where he had learned a combination of philosophical and experimental inquiry. Titchener was unconvinced of the value of the presumed objectivity attributed to scientific apparatuses themselves; instead, he believed that human beings' self-perceptions were central to psychological investigation. In fact, he maintained that "introspection is the one distinctively psychological method," and that "all objective data must, if they are to become psychological, be interpreted in the light of introspection." He contended that "the method of introspection, despite all attacks made upon it, is regarded by the great majority of present-day psychologists as the most important means of psychological knowledge."[83]

Despite Titchener's insistence on its popularity, a number of psychologists—including his own student, Ruckmick—were beginning to push introspection aside for more presumably objective methods. In his early studies, Ruckmick had often used introspective investigation as a way of understanding a subject's bodily experiences. In a study of the "kinaesthesis" of rhythm, Ruckmick's subjects reflected on the sensations they felt while listening to particular rhythmic patterns.[84] In a subsequent essay on smoking, Ruckmick asked subjects to record their bodily experiences while using tobacco, one of his subjects noting that "a slight tendency to belch was present, and the taste and smell of tobacco were prolonged."[85]

Even as Ruckmick was undertaking these introspective examinations, however, he was already beginning to question their validity. In an essay on the history of psychology written in 1912—the same year as Titchener's defense of introspection—Ruckmick opined that "the introspective method, peculiar to the psychologist, may offer a hindrance to the ready acceptance of the discipline."[86] Likewise, wary of emotions and increasingly concerned with scientific objectivity, Ruckmick began to fear that subjects' sensory experiences would distort their perceptions and make their introspections invalid. In an essay entitled "On Overlooking Familiar Objects," Ruckmick highlighted how people's personal experience can color their perception. Discussing perceptions of imagery, he stressed that humans show "an interpretative or perceptual attitude" that encourages them to distort visual memories.[87]

Ruckmick also became convinced that the physical concomitants of emotion took place beyond human perception. "Even when we have highly trained observers and bona fide emotions, like towering rage, there is frequently an absence in our reports of kinesthetic or organic sensations," he wrote in a letter discussing introspective analysis. "I have enraged students to the point that they said that they could almost kill me but even then there was no such report."[88] If people could never be clear about their own bodily sensations, then how could they be relied upon to adequately introspect?

This increasing distrust of introspection also served to push Ruck-mick and other psychologists further away from the philosophical perspectives on emotion that had been common in previous decades. While earlier psychologists had posed a variety of philosophical questions about passion and feelings—taking up issues of mind and spirit, in addition to inquiries about the brain—Ruckmick and his peers viewed such questions as counterproductive to a truly scientific psychology. While still a graduate student, after surveying a range of psychologists about their sense of the field, Ruckmick observed that "a number of these replies indicate, in no uncertain terms, that affiliation to philosophy is unfortunate; that if affiliation becomes necessary, . . . academic relationship with the biological sciences is preferred; and that the 'scientific approach' of experimental psychology is responsible for the steady progress of the discipline."[89] Years later, he would mention the "so-called psychologists . . . who elaborate upon a theory of consciousness, of mind-body relationship, or of emotion, who do not resort to experimental methods to substantiate their hypotheses."[90] In a review of Edwin Boring's book *A History of Experimental Psychology*, Ruck-mick similarly observed that Boring "believes that psychology should entirely relinquish 'its philosophical heritage'—with which the reviewer heartily agrees, in so far as any science can do so."[91]

Growing anxieties about introspection and philosophy were both products of, and contributors to, Ruckmick's and his peers' embrace of technological approaches to emotion. Seashore saw the emotions of music as buried in the subsensory rise and fall of the sound wave; only tools such as the tonoscope could capture music's true beauty. Ruckmick saw the human experience of emotions as equally superceptual. Introspection failed to provide scientific evidence because the subject really had no clear idea what was taking place when he or she felt an emotion. While Ruckmick and Dysinger collected introspective accounts of the film experience for *The Emotional Responses of Children to the Motion Picture Situation*, ultimately it was the psycho-galvanometer's measures of galvanic skin response that took center stage in their analysis, followed closely by the pneumo-cardiograph's measurements of pulse rate. In several cases, during the experiments for this monograph, researchers simply stopped taking introspective accounts because they found them difficult or unhelpful.[92] When they were taken, the researchers worried that they lacked "frankness and objectivity" and could "not be interpreted in terms of adequate scientific analysis."[93] As Ruckmick and his Payne Fund accomplices explained, "our trained observers [the Payne Fund study's term for the movie-watching subjects of their experiments][94] and some of the others recorded direct observations describing the type of emotion felt at certain points in the motion picture. But the main emphasis was placed on the amount of galvanometric deflection at various points in the film."[95]

Rather than dealing with subjects' perceptions, the psycho-galvanometer measured the electrical conductivity of their sweat glands. The eighteenth-century Italian scientist Luigi Galvani had made early strides in measuring these electrical currents. Galvani noticed that he could use electricity to make the muscles of a dissected frog's leg contract, first by applying electricity to the nerves of the leg, then by creating a spark near the leg while touching the nerve with a piece of metal, and finally by touching the nerve with a piece of metal without introducing any artificial current. In performing and testing these experiments, Galvani believed that he had proven the eighteenth-century theory of "animal electricity," which held that an electrical fluid flowed through the muscles and nerves of animals. This began a historic debate between Galvani and Alessandro Volta, who argued that the electricity Galvani noted was created by forces outside the body of the frog itself (for instance, in the atmosphere, or in the metal objects Galvani used in his experiments). Although the scientific community ultimately sided with Volta, "galvanism" left an important impression on the natural and social sciences.[96]

The "galvanometer" was originally created to monitor electrical currents applied to human muscles and organs for various medical reasons. Physicians of the late eighteenth and early nineteenth centuries believed that precise applications of electricity could cure such ailments as sciatica, rheumatism, herpes, constipation, and paralysis.[97] An 1802 article, "On Galvanism and Its Medical Applications," provided one of the first English descriptions of a galvanometer. The researcher hung a gold leaf and a wire in a glass tube. When current was sent to both the wire and the leaf, they formed an electrical attraction. Their degree of movement toward each other—captured on a brass scale attached to the apparatus—provided a measurement of the relative strength of the current.[98] During the middle of the nineteenth century, galvanometers were heavily developed and utilized in telegraphy, which also required careful monitoring of electrical charges. By the end of the century, a wide range of these instruments were in use throughout the sciences.

In the 1880s and 1890s, galvanometers began to be used in psychological research as scientists became interested in the connections between galvanic electricity and emotion. Psychologists of this period began to suggest that emotional states could be accompanied by changes in the electrical conductivity of the skin. By using a galvanometer to measure this electrical current, a researcher could determine a person's "psychogalvanic reflex." Charles Féré and Jean Tarchanoff were among the first to draw connections between psychogalvanic reflexes and emotions. The first decade of the twentieth century saw a still wider application of psychogalvanic measurements, including by Carl Seashore's mentor E. W. Scripture and the Swiss psychiatrist Carl Jung.[99] By attaching electrodes to a subject's skin and then monitoring his or her psychogalvanic reflex through a galvanometer, these researchers believed that they could monitor his or her accompanying emotional changes as well.

For the Payne Fund research, Ruckmick used a "Wechsler photographically recording galvanometer"[100] modified with equipment that he and other Iowa psychologists had been developing throughout the 1920s and early 1930s.[101] Ruckmick believed that one of the primary advances of these later psycho-galvanometers was their use of photographic technology to record the subject's galvanic deflection.[102] A researcher could measure the amount of electricity flowing through the subject's body and then record this on a psycho-galvanograph similar to the phono-photograph that Seashore had employed. Like Seashore's phonophotographic apparatus, the psycho-galvanometer also seemed to offer a stronger sense of emotional processes than subjects themselves could perceive. A chief benefit of measuring galvanic skin response, Ruckmick argued, was "the fact that it is not under voluntary control as is breathing, for example," which "eliminates errors initiated by the observer; i.e., he can have no direct control over the amount of deflection manifested by the galvanometer or other electrical registered device."[103] In this way, these pieces of equipment seemed to overcome the problems Ruckmick and other psychologists attributed to introspection. In Ruckmick's view, the psycho-galvanometer subverted a subject's perceptual deceptions to get to a deeper emotional truth of the body.

These developing laboratory apparatuses also helped psychologists concerned with overcoming their own emotional disturbances. As someone who celebrated impersonal writing and took issue with prose that stirred up emotional appeal, Ruckmick saw the psycho-galvanometer and pneumo-cardiograph as important tools of an emotionally restrained psychology. In the introduction to *Emotional Responses of Children to the Motion Picture Situation*, Dysinger and Ruckmick explained that "the affective life" had "for a long time resisted a direct frontal attack, partly because of traditional attitudes in regard to emotions." However, they continued, "the traditional attitudes are rapidly being dissolved through the impersonal approach of the psychological laboratory."[104] A few pages later, they stressed that although much of the public was passionate about the harmful or positive effects of movies, "as scientists we had no particular 'axe to grind.' We were simply

Figure 4.1

Ruckmick and Dysinger's motion picture viewing laboratory. The young boy's fingers are connected to the electrodes that run to the psycho-galvanometer on the table behind him. Reprinted from Wendell Dysinger and Christian Ruckmick, *The Emotional Responses of Children to the Motion Picture Situation* (New York: Macmillan, 1933), 16.

inquisitive and tried to get at the facts." As they put it, "the point here is made because even scientists sometimes lean toward certain theories and are, therefore, unavoidably prevented from reaching disinterested conclusions. Great care was taken to guard against any such possibility of criticism."[105] It was laboratory equipment such as the psycho-galvanometer that seemed to provide this protection.

The Iowa school's psychological researchers found a panacea in their newly utilized emotion-gauging apparatuses. Freeing themselves from emotional involvement, these researchers ascribed to their machinery the most empathetic access to their subjects' intimate feelings. The machines knew an individual's feelings better than the researcher, indeed, better than the feeling subject himself or herself, identifying emotional changes far too subtle for the subject to notice. By assigning to the apparatus the burdens of intimacy, scientists could frame themselves as distant bystanders, appropriately detached from their subjects. Even when scientists had to involve themselves with their subjects, this new approach suggested, it could be from a perspective of appropriate distance:

> During this period of preparation for the experiment, E made an effort, especially with children, to put O at ease by informal conversation. The procedure at this point was not equivalent from one O to another, since the point was not equivalent from one O to another. With a single exception, adequate rapport was established.[106]

Of course, the apparatus always established adequate rapport, going right to the deepest feelings of the subject under examination.

By casting the psycho-galvanometer in the role of empath, Ruckmick and Dysinger imagined their equipment as similar to its sister technology, the polygraph: as a path to the hidden emotional truth of their subjects (in fact, one of the inventors of the polygraph—John Larson—was at Iowa during Ruckmick and Dysinger's studies[107]). The term "polygraph" had been around since the late eighteenth century, when it referred to a writing apparatus used to produce two or more drawings or writings at once. In the late nineteenth century, the term became synonymous with any number of devices capable of measuring several different physiological processes at the same time. Only in the 1920s did the term become associated with the modern-day lie detector.

Despite these later elaborations, the eighteenth-century idea of duplication or reproduction adhered in the understanding of emotion-gauging apparatuses put forward by the media researchers of the 1920s and 1930s. These researchers saw the psycho-galvanometer and pneumo-cardiograph, like both the photograph and phonograph, as devices for capturing and replaying the emotional reactions of the subjects they studied. The electrical markings captured on the film of the psycho-galvanometer were a real inscription of the physiological activities of the media spectator, which were themselves a real result of that spectator's feelings toward the media he or she consumed. The researcher was a mere onlooker, magically freed from the burden of feeling, while the psycho-galvanometer was an empathetic medium, delicately "playing" the inner feelings of the spectator.

The Physicalist Audience

Five years after publishing *The Emotional Responses of Children to the Motion Picture Situation*, Ruckmick was still peddling the technological view of emotion that Iowa had helped to establish. A 1938 article in the *Washington Post* entitled "Hands as Lie Detectors" informed readers that Ruckmick had "invented an 'emotion meter,' with which, it is claimed, it is possible to measure the capacity of a person's anger, love, or faculty for telling untruths."[108] Nine months later, an article in the same paper explored Ruckmick's newest emotion meter—the dermohmograph—explaining that "its function is to smell out an emotion as surely as a terrier detects a rat":

> Reports of this work suggested to the University of Iowa students that the device might be able to show which boy a girl loves most, if the boys were to be paraded before her as she sat strapped to the machine. Ruckmick cautiously admitted that his dermohmograph might do the work, but declared "If I started that, I'd never have time for anything else."[109]

While this comical scene never came to pass, the particular view of emotions endorsed by Ruckmick's work for the Payne Fund and similar research of the 1920s and 1930s sheds an important light on the complex culture of emotion and technology that developed at this time.

First, the technologically driven media research of Ruckmick and his peers played into the rhetoric of the technological sublime discussed in the previous chapters. In *The Emotional Responses of Children to the Motion Picture Situation*, Dysinger and Ruckmick seemed to take the mythological power of technology for granted. While these two researchers believed that "the moving and talking pictures in their present vogue carry a tremendous sanction," they worried that "when the pictures are finally shown in color . . . and when the stereoscopic effect of tridimensional perception is added, . . . an irresistible presentation of reality will be consummated."[110] Technologically sophisticated movies were dangerous: "Profound mental and physiological effects of an emotional order are produced. The stimulus is inherently strong and undiluted by post-adolescent critical attitudes and accumulated and modifying experiences. Unnatural sophistication and premature bodily stimulation will result."[111] More sophisticated technologies promised still more sophisticated stimulations of emotion, violating the edict of emotional balance that Ruckmick and his peers held so dear.

The idea that the psycho-galvanometer could know an audience member's emotional experience depended precisely on this same sublime understanding of technology. In addition to his role as a founder of cinema, Auguste Lumière was praised by his contemporary scientists as an important innovator in "medical biology, pharmacodynamics, and experimental physiology."[112] As an important medical recording device for physiological movements, film is a cousin of the instruments of graphic inscription used in the Payne Fund studies. The same technological force that allowed cinema to visualize an "irresistible presentation of reality" allowed the phonophotograph and psycho-galvanometer to capture the hidden emotional truth in music and the body. Here, the Payne Fund studies offered a complicated relationship of technologies and sublimities, with two similar technologies pitted against and alongside each other. Motion pictures inscribed emotions in celluloid and then projected them for the consumption of their audiences. The psycho-galvanometer translated the audience members' physiological responses into numerical data, recorded on film. Positioned between these two technologies, movie audiences were passive collections of unperceivable bodily impulses waiting to be drawn out by the film or psychological apparatus.

The film historian Tom Gunning has suggested that a sense of the *uncanny* underlies many of our experiences with new technology.[113]

Ruckmick and Dysinger reacted against the uncanny emotional power that Münsterberg celebrated in the movies. However, motion picture technology still held a mystery and wonder for them. To a certain degree, Ruckmick, Dysinger, and their fellow technology-centered researchers maintained a sense of the "galvanism" that prefigured their experiments, not to mention the mesmerism that it helped produce. Even though Ruckmick became critical of the term "galvanic"—preferring the term "electrodermal response to galvanic skin response"[114]—his work maintained an implicit sense of animal electricity. There was a magical, uncanny relationship between movies and people that could be understood only through the scientific application of recording technologies themselves. The high-tech readings of the psycho-galvanometer had become the spirit photography of a presumably more enlightened time.

In this manner, Dysinger and Ruckmick elevated technology to a preeminent place in the media-audience relationship. Even audience members themselves were technologies. These researchers had essentially transformed the subjects of their study into the media through which these other technologies interacted; the audience was the electrical conduit through which the emotional meanings of the movie were translated into the transcriptions of the psycho-galvanometer. This view played into fears about the hypodermic effects of mass culture, particularly on children. Although certain older audience members benefited from an "adult discount" that prevented them from being emotionally carried away by a movie, adolescents were especially prone to excessive excitation. In any case, all audience members were equally electrical— their bodies humming along to a film's scenes in ways that allowed the psycho-galvanometer to chart their emotional peaks and valleys.

In reinforcing a sublime understanding of technology, Dysinger and Ruckmick's approach also helped to support the very commercial media system that the Payne Fund presumed to critique. In addition to their motion picture studies, in the 1920s and 1930s Payne Fund money went toward efforts by the National Committee on Education by Radio (NCER)—the chief proponent of nonprofit, educational radio—against corporate attempts to solidify the hold of commercial broadcasting in the United States.[115] However, the technologically focused research of Dysinger and Ruckmick affirmed many of the arguments supporting a commercial perspective on communication. For one, this study framed

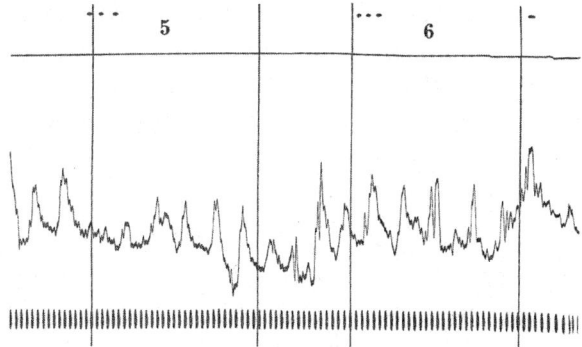

Figure 4.2

Photographic chart of the physiological responses of a nine-year-old boy watching the film *The Yellow Ticket*. Reprinted from Dysinger and Ruckmick, *Emotional Responses of Children to the Motion Picture Situation*, 94.

the moviegoing experience as a primarily transactional one. The technology of the film offered the spectator a visual product, which evoked an emotional response, which translated to a quantifiable, visual inscription. Much like Dale Carnegie's *How to Win Friends and Influence People*, these Payne Fund researchers suggested that emotions could be packaged and delivered, by the film to the viewer and from the viewer to the researcher. Emotions were but another commodity of exchange.

This symbiosis of scientific and commercial aims blurred the lines between academic and marketing research, as also happened with the stereoscope and speech apparatuses explored in the previous chapters. Motion pictures were both for-profit entertainment media and tools of social research. This meant that the researcher's laboratory equipment and research practices could quickly be adapted for commercial purposes. The Wonder Woman creator William Marston put his lie detector experience to work for Universal Pictures in analyzing the emotional impact of its films.[116] Carl Seashore's tests of music and emotion sold so well as measures of musical aptitude that when Ruckmick decided to move to Iowa, his mentor Titchener suggested that Seashore would likely cede control of the department so that he could devote more time to his commercial enterprises.[117] When Ruckmick left Iowa

in 1938—by one account because of controversies surrounding his dermohmograph[118]—he went to work for C. H. Stoelting, the same company that had manufactured Seashore's music tests and supplied much
of the equipment used by speech researchers at the University of Wisconsin and elsewhere.[119] Stoelting had also produced the Wechsler galvanometer that Ruckmick used in his own Payne Fund research.[120]

In addition to these academics and scientific corporations, the mainstream, commercial media companies were also helping to produce
the recording and measuring apparatuses employed in much of this
era's scientific research. Both Columbia and RCA Victor took a turn
at producing the records for Seashore's test.[121] Columbia also helped
to develop a technology that recorded heart sounds for the purposes
of playback and analysis.[122] G.E. had worked with medical researchers
to develop an electrocardiograph, a 1924 article in the *New York Times*
proclaiming "Electro-Cardiograph Developed by the General Electric
Company Renders Heart 'Voltage' Visible."[123] Cardiographs and similar recording instruments were especially well suited for both academic
and administrative purposes. They measured the *direct, individual
effects* that commercial media producers desired and that many social
researchers believed were the key to establishing a scientific practice of
media study.[124]

For the Payne Fund researchers, these scientific studies reinforced
commercial approaches to media in still more direct ways. When the
Payne Fund motion picture studies began, William Short compared
the motion picture industry to print publishing, asking a colleague to
"imagine that the art of Printing had only been recently invented, that
it had been immediately seized, commercialized and monopolized for
amusement purposes by a few men largely devoid of ideals."[125] Short
proposed a number of studies of the film industry's monopolistic practices, including the various ways that they influenced legislation and
exerted political pressure. The Payne Fund administrators, however,
worried that such criticisms of the industry would draw negative attention to the fund. Against Short's recommendations, they vetoed these
economic studies in favor of the more purely scientific studies that ultimately made up the Payne Fund monographs. Studies like Ruckmick
and Dysinger's were politically safe because they avoided questioning
the fundamental economic and political structure of the film industry.[126]

Even the authors of the more qualitative Payne Fund studies had trouble negotiating between social research on emotion, consumerism, and motion picture technology. W. W. Charters, the chairman of the studies, identified Herbert Blumer's book *Movies and Conduct* as the other book—after Ruckmick and Dysinger's—that took emotions as a primary concern.[127] Toward the end of the book, Blumer reprimands movie producers who might defend their films as "art for art's sake" without taking into consideration the values and wants of their audience:

> What may be intended by the producer and director as art, may be accepted by the movie public, or significant portions of it, as pornography. The difference, if it exists, is obviously a matter of interpretation. . . . What may evoke aesthetic satisfaction on their part may stimulate others in an unmistakenly contrary fashion. Unless the aesthetic values and interpretation of the movie public are changed to conform to those of the directing personnel, it is anomalous to defend commercial depictions on the basis of their art value, and to charge unfortunate effects to the basemindedness of people.[128]

Blumer here offers an interesting and vaguely populist attack on filmmakers. In the process, however, he also repeats the commercial exchange relationship these studies seemingly aimed to critique. That a filmmaker's responsibility is to the public's taste seems closely akin to saying that filmmakers should give the people what they want—which is quite similar to the argument that commercial broadcasters were making at this same time, in claiming that they were best capable of satisfying the needs and desires of mainstream audiences. In framing his concerns within the specific language of exchange, Blumer encountered the same problem inflecting Dysinger and Ruckmick's more technological approach.

In its focus on technology, Dysinger and Ruckmick's research inadvertently supported a larger argument in favor of the period's commercial broadcasters as well. Since 1928, the largely pro-commercial Federal Radio Commission (FRC) had interpreted its role as "bring[ing] about the best possible broadcasting reception conditions throughout the United States."[129] In the service of this goal, in 1928, the commission undertook a plan to reallocate radio licenses. General Order 40 forced 94 percent of broadcasters to change their radio frequencies, and "the 6 percent that were unaffected were chain owned or affiliated stations on clear channels."[130] Of the four features guiding the reallocation plan as explained by John Dellinger, the FRC's chief radio engineer, three dealt with technical interference issues. The forth addressed tensions between local, regional, and distant radio stations, but even this was couched primarily in terms of "heterodyne interference."[131] Focusing on clarity of transmission (as opposed to, for instance, more open access to the airwaves), the FRC interpreted broadcasters' service to the "public interest" in technological terms. The better a station's transmitter, the better the public interest was served. This played into the hands of the larger, commercially driven radio stations that could afford newer and more sophisticated equipment, forcing out many of the smaller, more diverse, noncommercial radio broadcasters with more limited transmissions. For the FRC commissioners—as with Dysinger and Ruckmick—a well-tuned apparatus was the fundamental factor in effective communication.

The ramifications of this technological take on emotion were intensified by Ruckmick's dismissal of philosophical modes of inquiry. Hugo Münsterberg's ruminations on such weighty topics as "the purpose of art" dealt directly with matters of aesthetics and ethics. Ruckmick and his colleagues eschewed such themes in favor of more tangible, scientifically verifiable claims. The Payne Fund scientists privately condemned Forman's moralizing summary of their work, preferring their studies to retain a more empirical presentation. Of course, such objections elided the more implicit moral claim, which suffused Ruckmick and Dysinger's work more generally, that emotional stimulation was bad for civilization and did harm to both audiences and researchers. In assigning the task of empathy to the psycho-galvanometer, Ruckmick and Dysinger upheld an ethic of emotional restraint even as they abdicated the moralizing role assumed by Münsterberg and other psychologists before them. For Ruckmick and Dysinger, the researcher's chief responsibility was to build a good apparatus and then let it fulfill its ethical duty: inscribe a subject's emotions on its sublimely technological surfaces.

The potential problems with this new, anti-philosophical perspective on emotion were palpable for the Iowa researchers. Among the research that took place in Seashore and Ruckmick's department were Wendell Johnson's late-1930s experiments with stuttering. One set of studies, completed as the MA thesis of Mary Tudor under Johnson's supervision, involved a group of orphan children. Central among her research questions, Tudor wanted to see whether telling stuttering children that they did not stutter and giving them positive affirmation of their "normal speech" could encourage them to stop stuttering. Likewise, she would test whether telling non-stuttering children that they did stutter, and warning them in various ways to be careful of their speech could induce them to stutter. Amid these various sorts of emotional feedback, the children's speech would be observed and—not surprisingly—phonographically recorded. Tudor ultimately found that children could not be induced to stutter through negative comments about their speech, though she did find that the non-stuttering children who were told that they stuttered became very self-conscious and inhibited in terms of their speaking. As with Ruckmick's own research, in Tudor's project, which years later became known as the "Monster Study," emotions were simply experimental variables rather than actual human experiences.[132]

Finally, and in line with the discussions of radio speech and research in the previous chapter, the understanding of emotion implicit in Ruckmick and Dysinger's film research took a decidedly white, middle-class, and narrowly American form. This is perhaps best demonstrated by the films they selected to test the emotional reactions of their research subjects. Ruckmick and Dysinger claimed that "the pictures used in the theater were selected almost at random."[133] Yet there was a great degree of similarity among these films, suggesting the sort of motion picture that both movie producers and Ruckmick and Dysinger believed would be emotionally stimulating for audiences. In discussing "The Feast of Ishtar," a short excerpt from the 1925 film *The Wanderer*, Ruckmick felt the need to explain that it "features extravagant scenes of oriental luxury." He noted specifically that depictions of "oriental dancing occur throughout."[134]

Themes of "orientalism" dominated the other film examples as well, which included the 1931 film *The Road to Singapore*. This film tells the story of an affair between a doctor's wife, Phillippa, and a well-known playboy, Hugh Dawltry, both of whom live as American expatriates in a coastal town in India. Calling attention to the presumably dangerous sexuality of this area, in one scene, Phillippa's husband, George March, says of his sister, "She's 18; that's a woman in the tropics." He adds that "this heat is bad enough on married women, but on young girls it's dynamite; makes them man crazy." Similarly linking native culture and heightened

Figure 4.3

An image from *The Wanderer* (Raoul Walsh, 1925) depicting the "oriental" scenes that Ruckmick and Dysinger believed were so emotionally stimulating.

sexuality, in another scene native drums beat as part of a nighttime festival to a goddess of love, as Hugh and Phillippa stare at each other longingly across a vast jungle of palm trees. In an earlier scene, Phillippa bumps into a person whom Ruckmick and Dysinger describe as "an ugly native."[135] According to their interpretation, "surprise and danger are clearly involved" in this collision.[136] Ruckmick and Dysinger labeled the vast majority of the movie's scenes either "dangerous" or "suggestive." Like the fictional George March, Ruckmick and Dysinger seemed to view "the tropics" as a source of great danger and sexuality.

Charlie Chan's Chance (1932), another of the films Ruckmick and Dysinger used in their study, also capitalized on "oriental" themes. The

movie takes place in New York City, rather than in Chan's native Hono-
lulu, but one of the two murder suspects whom Chan investigates with
the New York City police is nonetheless a "Chinaman" named Li Gung.
Although the groups they studied had higher average responses at
other moments in the film, Ruckmick and Dysinger identified a scene
in which Li Gung tries to kill Chan, but ends up killing himself by mis-
take, as the most "dangerous" scene of the film.[137] This likely reflected
Ruckmick and Dysinger's own sense of the perceived danger of this
presumably exotic confrontation. Though less explicitly, the other
two dramatic films used in the study, *The Yellow Ticket* (1931) and *His
Woman* (1931), also linked danger, sexuality, and exoticism. In *The Yellow Ticket*, a young Jewish woman is forced into prostitution in czarist
Russia, and *His Woman* begins in "Tamarind, a port in the Caribbean,"
where a freight ship captain played by Gary Cooper gets into a bar fight
after flirting with a native dancer. Echoing the beliefs of the companies
that produced stereoscopic tours of the world, Ruckmick and Dysinger's film selection suggested that there was something especially exciting about foreign or otherwise "oriental" scenes that made them perfect
material for studying the dangers of motion picture viewing.

The last two films used in the study, *Hop to It, Bell Hop* (1925) and *The
Iron Mule* (1925), were slapstick comedies used to measure the excitement
caused by humor. *The Iron Mule*—a parody of the 1924 film *The Iron
Horse*, which celebrated nineteenth-century locomotives—depended
precisely on the humor of a higher-technology audience looking back on
a lower-tech time. In one scene, the slow-running "Iron Mule" is passed
by a cow. In a later scene, it loses a tug-of-war with a horse that is tied to
it. Once the train gets moving again, it is attacked by a band of Native
Americans who unsuccessfully shower it with arrows—another "primitive" technology. One of the Native Americans chases a man from the
train, only to be rebuffed when the man hands him his "scalp," which
turns out to be a wig that reads "Sears and Roebuck" on the bottom—
a historical anachronism providing further evidence of the disjuncture
between different technological cultures. The subjects of Ruckmick and
Dysinger's study were expected to find a similar humor in the low-tech
scenes of *The Iron Mule* as they did danger and seduction in *The Road
to Singapore*. The emotions Ruckmick and Dysinger assumed they were
measuring—and that they ultimately hoped to control—were those of

the same white, narrowly American technocrats targeted by the Keystone and Underwood and Underwood stereoscope companies.

Transmitting Models of Communication

A headline from a 1933 article in the *New York Times* reporting on the Payne Fund studies went to the heart of the period's anxieties about film and emotion: "Overexcitement Is Seen."[138] Caught up in a culture of emotional control, the Payne Fund motion picture studies—and Dysinger and Ruckmick's work in particular—had sought to ferret out the emotional excesses of mass culture. Dysinger and Ruckmick had also taken aim at the excessive emotions of social science itself, pushing for the more quantitative, impersonal research that had taken hold in the early twentieth century. Drawing from Iowa's tradition of experimental psychology, Seashore, Ruckmick, and Dysinger approached the arts from a highly technological, empirical standpoint, seeing beauty and emotion as inscribed in the mechanical reproductions of the sound wave, motion picture, or psycho-galvanometer. As the first systematic study of media effects, the Payne Fund motion picture studies commanded a wide public audience, and Dysinger and Ruckmick's study, like the others of this collection, reflected a variety of cultural concerns about the new media back to the culture at large.

This technological view of media and emotion revealed a growing social scientific attitude that came to dominate American media research. Harold Lasswell made famous a conception of the communication situation that had originated at the time that he was employing the psycho-galvanometer in his own research: "Who says what in which channel to whom with what effect?"[139] This linear understanding of communication—and its more simplified "sender-message-receiver" variant—was based in the same thinking that drove Dysinger and Ruckmick's research; their technological experiments imagined a straight line from film to viewer in series with the wires of the psycho-galvanometer. It is difficult to imagine a better example of James Carey's "transmission model" of communication.[140] Work such as Dysinger and Ruckmick's helped to give this model its theoretical dominance, pushing aside a range of philosophical, ethical, and more broadly social questions in favor of a more narrowly individualistic conception of the media audience.

Likewise, the same technology-centric attitude toward media that had been deployed by the FRC was further institutionalized in the Communications Act of 1934. The opening of section 1 of the act, which describes the act's purposes, uses words such as "rapid," "effective," and "efficient," and focuses on the adequacy of broadcast "facilities." The first line mentions "the purpose of regulating interstate and foreign *commerce* in communication,"[141] suggesting the links between communication and the movement of goods (a regulatory attitude stretching back to the telegraph, which was governed in large part by railroad law). The word "transmission" appears no fewer than fifty times in the text of the act, while neither "community," "culture," nor "society" appears once (and although "public" does appear several times, it is usually in the limited sense of "public interest" that the FRC had established). The legislators who wrote this act were more concerned with the "transmission of energy by radio" than they were with the interactions among and between the members of the public. Like Dysinger and Ruckmick, the FRC (and the FCC after it) seemed to believe that if they committed themselves to technological questions, these other matters would somehow fall into place behind them. In the case of the 1934 act, this belief allowed broadcasting to become thoroughly ensconced in commercial interests, who largely pushed aside the amateurs and nonprofit visionaries who tried to oppose them.

Taking up the topic of emotional responses to motion pictures, Ruckmick and Dysinger had tackled a complex, politically and culturally charged topic. As several of my chapters demonstrate, recent scholarship in sociology, anthropology, and elsewhere has lamented the inattention to emotion in much social scientific thought, suggesting that emotions have remained a victim to their historical legacy as inferior to reason and therefore less worthy of research than more presumably serious topics. To their credit, Ruckmick and Dysinger sought to pay careful attention to emotional responses and to connect emotion to the larger cultural experience of moviegoing. Ultimately, however, Ruckmick and Dysinger's work reflected the highly troubled understanding of technology and emotion that predominated among many thinkers of their time. Seeming to share Leopold Stokowski's beliefs in the nearly psychic power of recording technology, Ruckmick and Dysinger worried over the power of cinema even as they took refuge in the powerfully etched celluloid of the psycho-galvanometer.

5

Connecting Centuries

The Legacies of Media Physicalism

An April 2010 *New York Times* article entitled "Digital Devices Deprive Brain of Needed Downtime" reported that

> at the University of California, San Francisco, scientists have found that when rats have a new experience, like exploring an unfamiliar area, their brains show new patterns of activity. But only when the rats take a break from their exploration do they process those patterns in a way that seems to create a persistent memory of the experience.[1]

The research the article mentions is that of Loren Frank, a UCSF scientist who had studied processes in the hippocampus of rats as they "replayed" various memories. According to the study, a rat's hippocampus "consistently replays past experiences during brief pauses in waking behavior."[2] It was up to the *New York Times* writer to substitute humans for rats and taking a break from one's digital device for "brief pauses in waking behavior," the assumption being that constantly talking on a cell phone is a bit like scurrying around like a rat.

The idea that digital technologies cause a kind of overexcitement has become a pervasive sentiment among contemporary Americans, if not Western culture more generally. On the one hand, this concern can be seen as a continuation of the tendency, at least as old as Socrates, to assume that new technologies bring new levels of emotional and intellectual stimulation. At the same time, the often explicitly physiological nature of these discussions of digital stimulation—as evidenced in this

New York Times article—has firm roots in the early twentieth-century ideas discussed in the previous chapters. In trying to understand the emotional effects of the new mass media, Carl Seashore, Christian Ruckmick, and others had sought to get "below" people's own experiences by employing various technologies that measured a range of bodily processes presumably impacted by media use. This approach adhered closely to the larger cultural understanding of early twentieth-century technology illustrated in discussions of motion pictures, radio, and the stereoscope. The new media technologies made emotions *tangible*, the argument went, allowing them to be captured and transmitted with a new kind of power. In the process, emotions themselves came to be seen as particular kinds of mechanical impulses—reflecting the apparatuses through which both media producers and academics attempted to take control of people's emotional lives.

Such were the assumptions of early twentieth-century media physicalism. Ruckmick believed that audiences ultimately did not have an authentic understanding of their own emotional responses to movies. Comprehending these responses required technologies such as the psycho-galvanometer, which gave a presumably objective recording of the biological processes indicative of true emotional stimulation. Seashore believed that he knew the beauty of music better than audiences or musicians themselves, because, he argued, beautiful music reduced to particular tonal vibrations that themselves produced pleasure in listeners. In a similar way, for both its champions and critics, the stereoscope created a set of sublime emotions because of the bodily experiences induced by its "three-dimensional effect."

This chapter begins by considering a collection of recent popular and academic discussions about media that reflect much of the rhetorical ecology of media physicalism demonstrated in the early twentieth century. Reminiscent of the debates that surrounded the emerging broadcast media, "the digital age" has experienced its own conflicted rhetorics of technological sublimity. A 1966 article in the *New York Times* suggested that the computer's ability to complete routine tasks "offers the chance to eliminate the drudgery that for millenniums has consumed the better part of man's time and energy." *Time* named the computer its 1982 "Machine of the Year," changing its "Person of the Year" award to reflect the technology's apparently revolutionary nature. The Internet and digital technology more generally have been celebrated by academics such as Henry Jenkins and Mark Poster and politicians as diverse as Al Gore and Newt Gingrich. Even the Vatican claimed that the new digital media could "contribute greatly to the enlargement and enrichment of men's minds and to the propagation and consolidation of the kingdom of God."[3] From the standpoint of these celebrants, the digital connections of the Internet seemed to promise the national or even global "heart" that had been associated with the telegraph.

As was the case with the new media of the early twentieth century, a number of popular and academic sources have been just as vocal about the destructiveness of digital technologies. The ill-fated computer HAL, of Kubrick's *2001*, offered one image of the devastating possibilities of digital technologies run amok. Written not quite fifteen years later, *Time*'s "Machine of the Year" article reported that Nils Nilsson, director of the Artificial Intelligence Center of SRI International, "believes the personal computer, like television, can 'greatly increase the forces of both good and evil.'" The media scholar Neil Postman similarly wrote that the computer had "usurped powers and enforced mind-sets that a fully attentive culture might have wished to deny it" (this built on Postman's earlier claim that, with television, human beings were "amusing themselves to death"). As early as 1994, a *Los Angeles Times* writer asked whether the Internet was becoming "a deep, dark addiction" for its most serious users.[4] Again echoing much of the sentiment of the early twentieth century, the same technological power that many believed would allow the Internet to bring people together into a newly unified community has led others to worry about the technology's destructive, addictive, hyperstimulating power.

Likewise, and similarly to the media of the early twentieth century, digital technologies have not only been consumed by the general public. Especially at the dawn of the twenty-first century, these technologies are also being employed by academics as they research the effects of digital technologies themselves. Like the studies of Christian Ruckmick, who used electrical film recording technologies to understand the emotional effects of films, much contemporary research is using digital technologies to make sense of the emotional effects of digital technologies. As it did in Ruckmick's time, this has made for a complex and often contradictory set of ideas about the power and impact of these technologies, especially among those who, like Ruckmick, are anxious about how the age's new technologies stimulate people's emotions. Here, the same sublime power that makes digital technologies damaging to a culture's emotional life seems to make them the ideal tool for measuring, recording, and controlling those emotions.

In tracing the complexities of these views of technology and emotion, this chapter offers a context for understanding the physicalism that pervades much twenty-first-century thinking about media, demonstrating how this era's thinking about communication technologies advance the narrowly technological view of emotion that came to prominence in the early twentieth century. This chapter begins by tracing a series of popular and academic discussions of the emotional effects of media that demonstrate contemporary media physicalism at work. The second section draws on arguments in the philosophy of mind to make sense of, and ultimately recast, the media physicalist understanding of emotion. The final sections consider some of the broader implications of media physicalist thought, including its particular ethics of media and the concrete ways it has become realized in American media policy.

Media Physicalism and the Technological Sublime

In approaching turn-of-the-twenty-first-century discussions about technology and emotion—in both their utopian and dystopian guises—the 1999 blockbuster film *The Matrix* offers a poignant entry point. Like *2001*, *The Matrix* offers a vision of computer technology gone wild. As a result of developments in artificial intelligence at the beginning of the twenty-first century, by the dawn of the twenty-second, humanity has been enslaved by a race of machines. Following a brutal war between humans and these new machines, humanity has been reduced to a series of batteries providing the power these computers need to operate. Now, human beings are grown and then placed in capsules that convert their body temperature into electrical power. "The Matrix" is a computer program—"a neural interactive simulation"—that creates a virtual reality to occupy the minds of these human batteries so they continue their energy production. As far as each human battery is concerned, he or she is living a normal human life in 1999—going to work, falling in love, and so forth. In *The Matrix*, HAL has won, and it is human beings who now "experience" computerized emotions.

Despite its extremely apocalyptic technological vision, the movie seems to suggest that it is ultimately technology itself that will free humanity. The heroes of the film are computer experts whose own technological mastery allows them to prevail over the Matrix. Neo, a young software engineer, is freed from his life as a battery by members of the resistance living outside the Matrix in the "real world" of circa 2199. After meeting members of the resistance "inside" the Matrix and taking a pill designed to arouse his real body, Neo awakens in his gelatinous battery-capsule. As insect-like machines fly by, a series of tubes are disconnected from Neo's body and he is transported to a ship operated by Morpheus, the leader of the resistance. On board the ship, Neo is again attached to a computer. He lies in a glass box, a series of metal objects probing his body, as images of his brain and internal organs flash on a monitor. If a computer has kept him immobile for his entire life, it will apparently be a computer that allows him to walk. Later, Neo "learns" a series of martial arts as a software program is uploaded into his brain. In a simulated fight sequence with Morpheus taking place in a specially designed computer program, both Neo and Morpheus demonstrate their mastery of their combat programming and their ability to work within and against the logic of the computer. Of course the elaborate, gravity-defying martial arts moves—some of the most celebrated of the movie—are made possible by the complex digital technologies on which the movie itself depends.

These ambivalent attitudes toward computer technology depended on the contradictory celebrations of, and concerns about, relationships between digital technology and human minds that characterized much of the 1990s. As Andy and Larry Wachowski, the two brothers who wrote and directed the film, explained in an interview, the movie was intended as a critique of "systems." "It's not just computers," Larry offered, "it's about anything you allow to think for you, systems of thought"[5] (this perspective would be borne out by the Wachowskis' explicit evocation of Jean Baudrillard's "simulacra"[6]). In offering an allegory of minds trapped by digital technologies, the movie advanced a conception of emotion very much in line with that of Christian Ruckmick and similar early twentieth-century psychologists. When Neo and Morpheus first enter the ship's simulated matrix together, Neo is confused by how real it feels to him. "This isn't real?" Neo asks Morpheus. "What is real? How do you define real?" Morpheus responds. "If you are talking about what you can feel, what you can smell, what you can taste and see, then real is simply electrical signals interpreted by your brain." If computers are taking over human reality, Ruckmick would likely agree with Morpheus, it is because they manipulate people's electro-emotional sensations with their own collection of electrical impulses.

If *The Matrix* offered a fictional account of bodies overtaken by computers through technological and emotional stimulation, a number of sources offered presumably real stories of these same developments. Anxieties about "Internet addiction," such as that expressed in the 1994 article from the *Los Angeles Times* mentioned above, made explicit references to the emotional power that digital technology held over computer users. The article quotes one user who apparently calls the Internet "my hallucinogen of choice." Sounding a bit like someone happy to be connected to the Matrix, she adds, "I love being able to slip into another body, another persona, another world." The writer takes pains to suggest that there might be good that comes from an assumed emotional freedom encouraged by the anonymity of cyberspace; however, the choice of interview subjects, all of whom the article claims spend "40 or more hours per week online"—which must have seemed like a large amount in 1994—clearly serves to highlight the potential intellectual and emotional damage of digital connectivity.[7]

These arguments have continued into the early twenty-first century. With reference to Twitter, Facebook, and similar recent technologies, a 2011 *Newsweek* article on "brain freeze" diagnosed the effects of information overload on the broader American and global culture. Citing a neuroscientist's measurements of activities in the brain, the article explained that with the increases in information caused by digital technologies, "frustration and anxiety soar: the brain's emotion regions . . . run as wild as toddlers on a sugar high."[8] An *NBC Nightly News* segment entitled "The Teenage Brain in the Digital Age," featuring Dr. Jay Giedd of the National Institute of Mental Health, alternates between shots of young people using digital devices and images of Giedd using a magnetic resonance imaging (MRI) system to measure their brain activity. While Giedd says that there is not yet sufficient evidence to indicate whether the increased activity caused by "information overload" is positive or negative, he is explicit that "there is a tax" on the brain as a result of such activities as multitasking.[9] Such reports seemed to demonstrate what *The*

Matrix had predicted: the takeover of human emotions—for better and worse—by various high-tech computer devices.

These ideas about technology and emotion have found still wider exposure, including in several best-selling books. Steven Johnson's *Everything Bad Is Good for You: How Today's Popular Culture Is Actually Making Us Smarter* purports to show how the increased activity and information of the digital age stimulates the brain and human emotions for the better.[10] In this book, Johnson, a science writer and journalist who has contributed to such publications as *Discover* and *Wired*, argues for what he calls "the sleeper effect": because of advances in technology, popular culture is getting more complicated and in the process creating more and better cognitive challenges for everyday people. As evidence, he cites such new media developments as the video game *SimCity*, whose game structure and organization—players need to consider a broad range of economic and political concerns as they build their own virtual city—require a complex set of mental activities. In fact, Johnson sees something positive about the "information overload" that many have criticized in the digital age. As he puts it, in comparison to the past, "the mind is more challenged mastering the dozens of new media forms—games, hypertext, instant messaging, TiVo—that constitute mainstream culture today."[11]

While Johnson's celebrations of information overload contradict much of the claims of Christian Ruckmick and his colleagues in the early twentieth century, much like Ruckmick, Johnson places physiology at the center of the relationship between media and the people who consume them. According to Johnson, in order to understand the emotional power of video games, "you need to look at game culture through the lens of neuroscience."[12] If the psycho-galvanometer provided this physiological perspective for Ruckmick, Johnson finds it in the form of the MRI and fMRI (functional magnetic resonance imaging). Likewise, and similarly to the Payne Fund's approach, Johnson highlights the essentially physiological relationship between media and emotions by making explicit links to drug use. Explaining his interest in the neuroscience of video games, Johnson writes that "if you're trying to figure out why cocaine is addictive you need a working model of what cocaine is and you need a working model of how the brain functions."[13] Seen through this perspective, media use becomes a series of pleasure- and reward-seeking activities, stimulating both the opioids, which are "the brain's pure pleasure drugs," and dopamine, which Johnson asserts triggers a "craving instinct" similar to that of drug addiction.[14] For Johnson, the physiological, drug-like effects of media pleasure explain "how video games get kids to learn without realizing that they're learning,"[15] and help establish the more mentally stimulating media experiences he argues are part of the digital age.

Nicholas Carr's book *The Shallows: What the Internet Is Doing to Our Brains*, another national best seller and a finalist for the Pulitzer Prize, takes a presumably opposite tack from that of Johnson.[16] A journalist who has written for such publications as the *Atlantic*, the *New York Times*, the *Wall Street Journal*, and *Wired*, Carr, like Johnson, has covered a variety of technological issues. If Johnson represents the optimistic side of the digital technological sublime, Carr certainly frames himself as a pessimist. For Carr, "the Shallows" describes how people interact with information in the digital age. Using himself as an example, Carr writes that "whether I'm online or not, my mind now expects to take in information the way the Net distributes it: in a swiftly moving stream of particles. Once I was a scuba diver in the sea of words. Now I zip along the surface like a guy on a jet ski."[17] According to Carr, search engines are among the many digital technologies contributing to this intellectual shallowness, as Carr claimed in a 2008 essay titled "Is Google Making Us Stupid?"[18] "Information overload has become a permanent affliction," Carr writes in *The Shallows*, "and our attempts to cure it," such as trying to organize it via Google, "just make it worse."[19] In contrast to Johnson, for Carr the speed and amount of information in the digital age are making for shallower intellectual and emotional connections across the culture at large. Carr is Walter Lippmann's "disenchanted man" for the computer era.

Despite the differences in the two writers' general evaluation of the digital age, the understanding of mediated emotion that underlies both Carr's and Johnson's claims is nearly identical. Early in his book, Carr explains that "media work their magic, or their mischief, *on the nervous system itself*," making explicit his own commitments to a physiological perspective.[20] According to Carr, current neuroscience shows that, at their most fundamental level, human experiences reduce to a set of neurological transmissions. "Thoughts, memories, emotions," he writes, "all emerge from the electrochemical interactions of neurons, mediated by synapses."[21] In the case of media, human experience develops at the nexus of a particular medium's technological properties and the neurological processes it naturally creates. For instance, because books require "deep, attentive reading," Carr explains, people in a culture where books predominate will have brains with neural pathways programmed for attentiveness. Because of the neurological features of book reading, even "cruder, crasser, and more trifling works" are more intellectually stimulating than the contemporary digital media that Carr criticizes. "Whether a person is immersed in a bodice ripper or a Psalter," Carr argues, "the synaptic effects are largely the same,"[22] and it is these neural effects that give us the true picture of the book's impact on a person. For Carr and Johnson alike, we know the importance and effect of a media technology when we understand how it interacts with the neurochemistry of the brain.

Owing to this focus, Johnson and Carr both celebrate a range of medical technologies used for brain mapping. For his 2004 book, *Mind Wide Open: Your Brain and the Neuroscience of Everyday Life*, Johnson had subjected *himself* to analysis though a range of neuroscientific equipment.[23] The book opens with Johnson attached to a "biofeedback system" much like the psycho-galvanometer Ruckmick used in his research. "Because damp skin conducts electricity more efficiently than dry skin," Johnson explains, "the electrodes on my palms could track how much I was sweating by monitoring changes in conductivity over time."[24] Once Johnson is allowed to see the data that result, he imagines that he is experiencing himself in an entirely new way: "I looked at that paper and thought: I've caught a glimpse of *me* here, viewed from an angle that I've never experienced before."[25] Like Christian Ruckmick, Johnson assumes that the flow of psychogalvanic electricity allows access to a reality of the body that is unavailable without it.

Johnson's *Everything Bad Is Good for You* and Carr's *Shallows* both hold to this basic premise, assuming that fMRIs and similar technologies tell us something about the human experience of media that is invisible without them. The evidence both writers provide for how digital technologies "rewire" the brain—for the better for Johnson, and for the worse for Carr—comes from a range of scientific studies employing new brain mapping techniques. Stressing the technological sophistication of this neuroscientific equipment and its relationship to other digital technologies, Johnson writes that "the age of brain imaging, genome mapping, and the microchip stacks up nicely against past eras—particularly when you look at the sheer number of individuals contributing groundbreaking work as opposed to the isolated geniuses of the past."[26] Despite his proclaimed pessimism about digital culture, Carr is a strong advocate of digital brain imaging as well. His celebration of reading, for instance, relies to a large extent on evidence provided by using "brain scans to examine what happens inside people's heads as they read stories."[27] Likewise, it is ultimately digital brain scans that demonstrate for Carr the shallowness of people's thinking as they use Google and other Internet sources. The same technological sophistication that presumably allows video games and Google to rewire human brains guarantees for both Johnson and Carr the fidelity of the high-tech, digital brain images that buttress their arguments.

As Ruckmick's Payne Fund research demonstrated, using a technology to study itself is bound to result in a host of complications and contradictions. It is these contradictions that in large part allow Johnson and Carr to come to such disparate conclusions using nearly identical information and approaches. For instance, both Johnson and Carr discuss the same 2000 study of the hippocampi of London taxi drivers. The researchers in the study, neuroscientists in London, used MRIs to compare the hippocampi of cab drivers to non–cab drivers, ultimately finding that "the posterior hippocampi of taxi drivers were significantly larger relative to those of control subjects." They reason that the mental work that goes into navigating a complex city increases the size of the posterior hippocampus because it "stores a spatial representation of the environment and can expand regionally to accommodate elaboration of this representation in people with a high dependence on navigational skills."[28]

For Johnson, this study provides evidence of the capacity for the brain to grow when new information is acquired. "This is the magic of the brain's plasticity," he writes of the study: "by executing a certain cognitive function again and again, you recruit more neurons to participate in a task."[29] Johnson draws parallels between this cab driving situation and other real-life interactions, as well as with the interactions we experience on television, online, and in other media. In his view, the complex "social networks" depicted on reality television or built on places like Facebook require their own complicated mental maps. Presumably, an especially active Facebook user would have a posterior hippocampus like a veteran cabbie. In contrast, Carr focuses on the potentially negative effects he sees implied in this study, highlighting that while the posterior hippocampus was larger in the cab drivers studied, the anterior hippocampus was smaller. Because, as Carr highlights, "the shrinking of the anterior hippocampus might have reduced cabbies' aptitude for certain other memorization," Carr sees the neural changes demonstrated in the study as evidence for just the brain damage he sees accompanying new media. "The constant spatial processing required to navigate London's intricate road system" has, for Carr, a parallel in the constant information processing of digital culture.[30]

These two interpretations of the London taxi study tell us much more about contemporary cultural attitudes toward technology than they do any actual effects of technology on the brain. If this study had been conducted in the early twentieth century, when the automobile was the new technology of the age, car critics would no doubt have decried the enlarged hippocampus as evidence of the corrupting influences of "speed mania." To those for whom the automobile is old hat, its status as a technology is easy to ignore, but it is here where Johnson's and Carr's discussions begin to fall apart. If both alter one's brain wiring, then is driving a cab better or worse for you than reading a book? Would people who want to increase their brain's capacity be better served by surfing the web, or navigating the streets of London in a hackney carriage? Are frequent drivers smarter than frequent web surfers, or are both somehow mentally incapacitated? Where are the calls for more—or less—driving? If these questions cannot be answered—the researchers of the London taxi driver study do not themselves take a position on the relative goodness of enlarged hippocampi—then how can we extrapolate between cab driving and web surfing or judge the relative worth of "neural pathways" in general? What does this say about the physicalist notion of media experience more broadly?

Popular works such as Johnson's and Carr's demonstrate just how widespread arguments about the centrality of physiology to media experience have become; these claims, and the sublime rhetoric on which they depend, are not confined to journalists, however. In an introduction to brain imaging in the journal *Media Psychology*, a group of researchers argues that the "ability to watch the brain and its responses to various media material in real time" opened up by fMRI imaging "promises the potential of finding biological bases for the behavioral changes from media exposure that have been observed for decades."[31] It is just this sort of evidence that Gary Small, the director of UCLA's Memory and Aging Research Center, and Gigi Vorgan attempt to provide in their book *iBrain: Surviving the Technological Alteration of the Modern Mind*.[32] Small and Vorgan, whose studies provided some of the evidence for Carr's claims in *The Shallows*, use fMRI equipment to measure the brain activity of people using a range of digital technologies. In the process, they provide neurological evidence for such conditions as information overload (what they call variously "techno-brain burnout" and "brain strain") and the generation gap between "digital natives" (those who have grown up using digital technologies) and the digitally naïve. In each case, the images of the fMRI become the authority on people's relationships to digital technology.

The strength of fMRI research for Small and Vorgan rests in its ability to locate and visualize various processes in the brain. By showing precisely what portions of the brain are stimulated when searching Google, they argue, we get a better sense of how computer users experience the process and what it means for them mentally. Exploring a different digital experience, they offer that "computer games depicting violent scenes activate the amygdala."[33] In another fMRI experiment they discuss, "areas in the frontal lobe of the brain that control positive emotions—the insula and the anterior cingulated—lit up in response to a strong car brand, Volkswagen, but not to a lesser known brand, Seat." From this research, Small and Vorgan conclude that "our brains are hard-wired to seek out established brands."[34] If people decide to buy one of these cars, it is presumably because it has triggered their "shopping instinct," which fMRI research shows us is driven by "the dopamine-rich area" of "the nucleus accumbens."[35]

As it did for Johnson and Carr, the contemporary rhetoric of the technological sublime pervades much of Small and Vorgan's discussion. Like Ruckmick, Lippmann, and others in the early twentieth century, Small and Vorgan assume that there is something particularly powerful about the digital age that sets it apart from other moments in time. "The printing press, electricity, telephone, automobile, and air travel were all technological innovations that greatly affected our lifestyles and our brains in the twentieth century," they write, drawing together several non–twentieth-century inventions in the process. "However, today's technological and digital progress is likely causing our brains to evolve at an unprecedented pace."[36] It is this same progress that presumably gives strength to Small and Vorgan's high-technology brain imaging research and makes these studies so persuasive for Johnson and Carr.

In line with this sense of technological progress, for Small and Vorgan there is something both technologically inferior and romantically quaint about predigital technologies. For instance, they lament that with the presumed speeding up of our culture, "even the traditional party invitation is being replaced by the e-vite," assuming perhaps that there is something about a paper invitation that makes it better for the brain than an electronic one. The "traditional love note" is also a casualty of the digital age, but here the problem is less the note's precise form and more the potentially public nature of its communications. In an age when "love notes" are more likely to appear on someone's Facebook page than in his or her mailbox, Small and Vorgan argue that the ritualized intimacy of these notes—and thus a particular kind of mental connection—is largely lost. However, as I point out in chapter 1, at earlier points in time love notes were distinctly public—sent to the family of one's lover rather than the lover himself or herself. Was this moment more like the digital age as a result? What is precisely the right amount of intimacy for the brain or the heart, and how do we know?

Small and Vorgan's arguments largely depend on imposing a contemporary notion of the technological sublime on all the various technologies they address. "Old media" are never as stimulating as "new media." A "traditional love note," however ill defined, must have a more natural effect on the brain precisely because it is "traditional." By the same token, if the computer—a new technology proclaimed to have great powers over the brain—stimulates the amygdala, then there *must* be something dangerous, empowering, or, at the very least, important about amygdala stimulation. The power of brain imaging derives from the same source. For Christian Ruckmick, the only way to match the technological power of a motion picture was with motion picture–driven research technology; brain imaging research on digital technologies likewise pits one powerful digital technology against another. The fMRI has a magical quality that Sir David Brewster would surely have envied.

Kelly Joyce has explored the mythologies of truth that surround MRI equipment within both the medical and science fields and the wider culture. Like Brewster's stereoscope, the MRI derives its authority from the assumed power of visualization. If there is a technological truth ascribed to photographic images, Joyce reasons, "then MRI as a high-status machine that generates an entire series of pictures must produce even more accurate and certain knowledge."[37] The great expense of MRI equipment contributes to this mythological power. Each machine's high cost highlights its assumed scientific sophistication even as the need to pay for it creates an impetus for putting it to use. Over the past decade, MRI use, like the faith in its technological truth, has risen considerably. As a result, in the medical field, healthcare costs have increased and doctors have begun to spend less time with their patients since they can pass along to the MRI many of the diagnoses they once performed directly on the patient.

Like Joyce, the psychologist William Uttal has argued that because of the "recent development of computerized tools" that allow scientists to explore the central nervous system, "the brain seems to have plunged off the rock of scientific certainty into a lake of unknown with an exuberance typical of a science suddenly provided a powerful new tool—or, perhaps, of a child given a new toy."[38] Whereas Joyce is a sociologist concerned with the larger economic and cultural factors contributing to the MRI's mythic status, as a scientist himself, Uttal calls into question a number of the scientific assumptions on which psychologically centered MRI studies are based. To what extent, he asks, can various psychological processes be "localized" in the brain, when the psychological processes are themselves often—and necessarily—poorly defined? According to Uttal, even so basic a process as "looking at something"— obviously an important component of media experience—"is fraught with technical and conceptual difficulties of enormous proportions."[39] What about more complex processes, such as deciding to buy something (Small and Vorgan's "shopping instinct"), feeling pleasure about a name brand, or reading a Wikipedia page?

The high-tech status of digital brain imaging makes it all too easy to ignore these conceptual problems and contradictions. The desire to use MRI equipment to quantify and order the processes of the brain is driven by the same "cultural logic of computation" that Carr, in his criticisms of Google, claims to be against.[40] In their faith in brain imaging, Carr, Johnson, and Small and Vorgan all advance this logic of computation in ways that have important cultural implications. For one, they all

seem to buy into the idea that cultures with more developed technologies will be smarter or more advanced than those without them, reiterating the kind of argument that Keystone View Company and Underwood
and Underwood attempted to sell alongside their stereoscopes. Indeed,
if digital technologies actively "rewire" people's brains, then computer
users will have brains that are substantially different from those of non–
computer users. In his celebratory claims that video games and similar
technologies are making people smarter, Johnson is explicit that digital
brains have been rewired for the better. People who use digital technologies, he argues, have higher IQs and are stronger at "problem solving,
abstract reasoning, pattern recognition, spatial logic."[41]

Carr shares this sense of technological progress, but places its pinnacle
at a different period in history: with the invention of the book. Here, he
reiterates a common argument that book literacy offered a special mental
stimulation that was not possible without it. His notion of the book as
a technology, however, is itself largely a modern fantasy. Unlike digital
technologies, he argues, books are not fragmented; they are consumed
as deep, coherent wholes rather than mere snippets of information.
Similarly, whereas the Internet is inherently multi-mediated and tied to
larger networks that distract people with a morass of information, books
require a singular attention and individual involvement. In the eighteenth
and nineteenth centuries, however, books were regularly fragmented and
shared. Readers copied quotes into their "commonplace books" and sent
along passages, or pages, to others with their letters. Thomas Jefferson
famously created his own Bible by literally cutting out passages he did not
believe Jesus said, and reassembling what was left into his own version of
the gospels. In celebrating "the book," Carr holds up a narrowly middle-
class ideal of intellectual and technological attainment—a lone individual
focusing his or her attention on a singular work—that may or may not
capture what "reading" has meant for most of its history (and certainly
reading has very rarely taken place inside an MRI tube).[42]

The fact that Carr locates this heightened moment of technological
civilization in the past does not prevent him from championing a larger
myth of technological progress, even if he sometimes views this progress negatively. For Carr, "newer" technologies that are farther down the
chain of technological progress are simply "brainier" than others: they
affect the brain for better (books) and worse (Google) with increasing

sophistication and power. As a result, other technologies, and their users, seem stuck in some primitive technological wasteland. "When a ditch digger trades his shovel for a backhoe," Carr explains, "his arm muscles weaken even as his efficiency increases. A similar trade-off may well take place as we automate the mind."[43] Even from within Carr's argument it would seem necessary for the ditch digger's transition to have an impact on his brain—perhaps along the line of what would happen if he started driving a cab. That Carr doesn't immediately assume so suggests the relatively low-tech status of both the ditch digger and his tools on Carr's scale of technological progress.

Perhaps even more so than Johnson and Carr, Small and Vorgan offer a vision of technological progress that would make Christian Ruckmick blush. Small and Vorgan suggest that new technologies play a fundamental role in evolution, which they say "essentially means change from a primitive to a more specialized or advanced state." This involves both brains and technologies and is an ongoing process: "When your teenage daughter learns to upload her new iPod while IM'ing on her laptop, talking on her cell phone, and reviewing her science notes, her brain adapts to a more advanced state by cranking out neurotransmitters, sprouting dendrites, and shaping new synapses."[44] Here, the so-called information overload of the digital age leads to a more advanced brain state, apparently placing the modern texting teenager high on the evolutionary scale. Given the mythology that surrounds it, the fact that a culture has MRI and similar brain-scanning technologies would presumably contribute to its advanced evolutionary position as well, providing people with newer and better ways of understanding themselves that lower-technology cultures do not have. In these ways, for those who see new technologies as part of an overall "advancing" or "speeding up" of culture, lower-technology cultures are inevitably slow—understood variously as relaxed, organic, quaint, primitive, non–self-aware, non-evolved, and dim-witted.

If Small and Vorgan's arguments provide evidence for the mental superiority of people with new technologies, they also offer supposed proof of the mental differences between men and women. "Evolution has programmed men and women to behave differently," they write. They explain that "men's brains are hard-wired to focus on small details and to grasp visual and spatial concepts more readily"—voicing the sort of claim that lead many in the early twentieth century to suggest that men were more technologically savvy than women. Offering a stereo-typically gendered notion of emotion as well, Small and Vorgan write that women's brains are "hard-wired" to "experience more empathy," whereas the male brain "tends to operate with greater emotional detach-ment."[45] Such claims ignore much of the history of gender and emo-tion, such as the periods when men have been encouraged to embrace their emotions (as I discuss in chapter 1). In both of the above cases, the empirical data of the brain scan—that the brain changes when people use different technologies and that differences can be found between male and female brains—becomes a scientific truth supporting long-standing ideological arguments about civilization and gender.

Finally, as was the case with much early twentieth-century media research, these neurological studies are inevitably *administrative* in ori-entation. Because they seek a direct, causal relationship between media and people, these studies ask the questions of most interest to propagan-dists, marketers, and others who seek to use media messages for various kinds of influence. In the early twentieth century, these links were clear-est with those scholars who went to work for the media industry or, like Carl Seashore, Smiley Blanton, E. H. Gardner, and E. Ray Skinner, made their own commercially available products. Digital research on technol-ogy and emotion has created similar connections and applications, as the work of Rosalind Picard illustrates. A professor in MIT's famed Media Lab, Picard, in her 1997 book *Affective Computing*, discussed the possibil-ity of creating computers that both represented and understood human displays of emotion.[46] Among other things, she offered that if people were fitted with devices that measured their physiological responses as they completed various computer tasks, those computers could adapt to the basic emotions of each individual user. Such a computer could monitor a person's "cognitive load" to mitigate the assumed stresses of informa-tion overload (for instance, by controlling when certain kinds of e-mails were delivered).[47] Like Johnson, Carr, and Small and Vorgan, Picard was suggesting that computerized emotional sensors could offer a solution for the emotional overstimulation of the computer age.

The marketing implications of this technology were not lost on Picard. Since writing the book, she has helped to found Affectiva, a company that creates computer technologies intended to help market-ers better understand and target consumer emotions.[48] For instance, the

Affectiva Q Sensor is a wearable device that monitors a person's body temperature, movement, and skin conductivity—making it essentially a digital version of the psycho-galvanometer used in Ruckmick's motion picture studies. In fact, a video promoting the sensor shows footage of a young boy's physiological responses as he watches a trailer for an Alvin and the Chipmunks movie, suggesting the commercial uses of the technology within the media industry. Another Affectiva product, Affdex, is claimed to allow computers to read people's emotions by interpreting their facial expressions via a webcam. As viewers look at various advertisements or products on their computers, this technology analyzes their emotions in order "to give marketers faster, more accurate insight into consumer response to brands and media."[49] Even more explicitly than in Ruckmick's work, the goal of this emotion-measuring equipment is high-tech empathy on behalf of better marketing.

None of these observations should suggest that neuroscientific research should not be done, or that it should not be done in studies of technology or emotion. Looking at the inner workings of the brain gives us important information, just as studying other biological processes does. Likewise, there is nothing inherently wrong with administrative research. However, administrative research on media has generally eschewed questions about history, politics, and economics in favor of more limited questions about the bodies' immediate physiology or behavior. In the case of studies of emotion, this is all the more pronounced and is tied to early twentieth-century anxieties about the emotional life and the status of those social sciences that research it. Brain imaging and similar physiological studies of mediated emotion are seductive for researchers and the general culture for how they frame our relationship with technology through the auras of quantification, science, and technology itself.

What gets left out in the process is not only the larger context of media relationships that may be equally or more important, but the larger view of emotion and technology as historically contingent concepts within our thinking about media itself. Without this historical awareness, media analyses—both academic and popular—can too simply reinforce contemporary ideological assumptions about media and emotion without asking how those assumptions might be shaping the discussion in the first place. How do we separate the technologies used for studying media from media as technologies themselves? How do we distinguish the rhetoric of technology that surrounds both from presumably empirical claims about their power? How does this thinking and talking about media affect the media situation itself, including the experience of the audience? Such questions require greater historical and philosophical reflection than contemporary media physicalism has tended to undertake.

A Chicken, a Bat, and a Neuroscientist Named Mary Walk into an Internet Café

In *Mind Wide Open*, Steven Johnson relates an odd-sounding study from the 1990s. According to Johnson, the neurophysiologist Jaak Panksepp had used emotion-measuring technologies to demonstrate that chickens liked certain kinds of music better than others; they had apparently expressed a particular preference for Pink Floyd. Johnson was actually conflating two of Panksepp's studies. He had published research showing *human reactions* to Pink Floyd as well as a study demonstrating chickens' general physiological responses to music.[50] Johnson's eagerness to endow chickens with specific media emotions makes sense in the context of his larger commitments to a physiological perspective on media. If the final truth is in the physiology, then people cannot be trusted to tell us their real emotional responses any more than chickens can. The real authority, for both person and chicken, is the emotion-measuring machine.

This is just the sort of thinking that Thomas Nagel was addressing in his well-known 1974 essay "What Is It Like to Be a Bat?"[51] He posed the central question of the essay in order to address a wave of physicalism he saw becoming dominant in philosophy. As he explained in an earlier essay, physicalists subscribed to "the thesis that a person, with all his psychological attributes, is nothing over and above his body, with all its physical attributes."[52] In Nagel's bat essay, among those with whom he associates physicalism's "recent wave of reductionist euphoria" is Daniel Dennett, whose scientific explanation of consciousness Nagel had already called into question.[53] In his book *Content and Consciousness*, Dennett had argued for a philosophical approach driven by both scientific brain research and digital computing. "How are commonplace observations about thinking, believing, feeling pain to be mapped on the discoveries of cybernetics or neurophysiology?" Dennett asks at the book's opening.[54] Although Dennett suggests that his goal is to maintain the links between personal (how a person imagines himself or herself to feel) and sub-personal (that person's brain states) levels of experience, he ultimately argues that it is the neural state that is the fundamental condition of human consciousness. If we cannot easily map our human terms for feelings (love, pain, desire, and so forth) onto particular brain states, argues Dennett, it is because our verbal expressions "are not the ultimate vehicles of meaning, for they have meaning only in so far as they are the ploys of ultimately non-linguistic systems."[55] If we want to understand what is *really going on* in people's consciousness, we shouldn't ask them. We should explore the deeper truth of their non-linguistic systems of neurochemistry.

In "What Is It Like to Be a Bat?," Nagel critiques this argument by highlighting discrepancies between knowing about brain states and knowing about experiences. For *consciousness* to exist in any creature, writes Nagel, there must be "something there is like to *be* that organism"—something that the organism can feel and experience.[56] Unless the organism has a sense of what it is like to *be* itself, in whatever form that takes, Nagel would argue that it is not truly "conscious." He calls this the "subjective character of experience." Nagel takes it for granted that any number of animals will have this kind of consciousness, but selects bats as his primary example. If bats are conscious, then there must be *something there is like* to see as a bat sees—a subjective character to the experience of bat vision. Of course, this is not "vision" in any sense that human beings know it, since bats "perceive the external world primarily by sonar or echolocation, detecting the reflections, from objects within range, of their own rapid, subtly modulating, high-frequency shrieks."[57]

For Nagel, human beings can never know what it is like to be a bat because they can never have the subjective experience of echolocation or any of the other components of bat experience. Although we can try to imagine what it would be like to navigate the world via sonar, or think about what it would feel like to hang upside down in the dark, bats' subjective experiences will remain out of reach. "In so far as I can imagine this (which is not very far)," writes Nagel, "it tells me only what it would be like for me to behave as a bat behaves. But that is not the question. I want to know what it is like for a *bat* to be a bat. Yet if I try to imagine this, I am restricted to the resources of my own mind, and those resources are inadequate to the task."[58] In considering bat experience in this manner, Nagel ultimately highlights the subjective character of human experience as well; after all, a bat philosopher would have no better luck imagining what it would be like to be a human. The same goes for human and bat neuroscientists who would presumably reduce experience to a set of "objective" neurological measures. "After all," asks Nagel, "what would be left of what it was like to be a bat if one removed the viewpoint of the bat?"[59] By calling attention to the subjective character of experience, Nagel highlighted the portion of experience that cannot be reduced to neurophysiology or brain states—both for bats and people.

In "Epiphenomenal Qualia," another important essay in the philosophy of mind, Frank Jackson offered a further critique of physicalism's neurological understanding of human experience.[60] Jackson builds his argument around several thought experiments. The most notorious involves Mary, who has spent her life in a room that is, by some quirk of lighting, completely absent of color. Everything she sees, including herself, appears in black and white. Over the course of her secluded life, however, Mary becomes an expert in the neuroscience of vision, including the neuroscientific bases of color perception. She learns all of this through a black-and-white television set that shows lectures and other information on the neuroscience of the brain. One day she is finally able to leave her room, after which she sees a *red* flower for the first time. When she does so, asks Jackson, does she learn something new about the experience of color? If so, he argues, then physicalism is false.

Jackson calls this the "knowledge argument," and he claims that when Mary emerges from her room, "it seems just obvious that she will learn something about the world and our visual experience of it."[61] His argument rests in large part on his commitment to "qualia," which he defines as "certain features of the bodily sensations especially, but also of certain perceptual experiences, which no amount of purely physical information includes":

> Tell me everything physical there is to tell about what is going on in a living brain, the kind of states, their functional role, their relation to what goes on at other times and in other brains, and so on and so forth, and be I as clever as can be in fitting it all together, you won't have told me about the hurtfulness of pains, the itchiness of itches, pangs of jealousy,

or about the characteristic experience of tasting a lemon, smelling a rose, hearing a loud noise or seeing the sky.[62]

Although Jackson specifically distinguishes his argument from Nagel's "What is it like . . . ?" argument, the resulting conclusion is largely the same. There is an experience of seeing red, a particular set of qualia, that Mary cannot know before she has seen red. It is these qualia, like the itchiness of itches, for which physicalism cannot account. To put it another way, a physiologist might be able to express precisely what would happen to someone's hand if an anvil fell on it, but that doesn't mean that they have full knowledge of the experience (at least hopefully!). It seems obvious that once the anvil hits, they, like Mary, will find their neurological knowledge insufficient in some fundamental ways.

These phenomenological arguments against physicalism have seen various sorts of critiques and extensions (and Jackson now includes himself among the critics of his own thought experiment).[63] Not surprisingly, Daniel Dennett has been one of the chief defenders of physicalism against these challenges. In his own thought experiment, Dennett discusses RoboMary, a robot who has black-and-white cameras for eyes.[64] In Dennett's argument, RoboMary is able to know what it is like to see red nonetheless—just as Dennett presumes Mary could. After analyzing the internal processes of other robots that *are* equipped with color cameras, RoboMary writes an algorithm to transform the images she sees into their color equivalent. Presumably Dennett assumes that neuroscientists can do just this; in prying open the brain, they can offer a picture of the internal algorithms that produce *the feeling of redness*, and thus make it open for reflection. In a response to both Nagel and Jackson, Paul Churchland similarly suggests that just as musicians can learn to make complex discernments between different musical pitches, if Mary is truly an *expert* in neuroscience, she would have a knowledge of herself and the physical world that would allow her to expertly imagine color. She would "not identify her visual sensations crudely as 'a sensation-of-black.'" Rather, she would identify "them more revealingly as *various spiking frequencies in the nth layer of the occipital cortex.*"[65] For both Dennett and Churchland, neuroscience gives people the full power of the emotion-sensing machinery on which it depends. No less than RoboMary, Churchland's Mary is a walking MRI.

David Chalmers has offered responses to these and other presumed refutations of the knowledge argument. As Chalmers explains, Churchland's case depends on an assumption that "various spiking frequencies in the nth layer of the occipital cortex" are equivalent to the feeling of redness in the same way that water is equivalent to H2o. Even if someone does not know that water and H2o are the same, everything he or she knows about water holds for H2o as well. Similarly, the physicalist argument goes, everything Mary knows about the spikes in her occipital cortex is true for the feeling of redness even if she has not yet directly connected the two. Chalmers argues, however, that "whenever one knows a fact under one mode or presentation [e.g., Superman can fly] but not under another [e.g., Clark Kent can fly], there will always be a different fact that one lacks knowledge of—a fact that connects the two modes of presentation [e.g., that Clark Kent *is* Superman]."[66] According to Chalmers, if Mary has not yet experienced "the feeling of redness," she is missing an essential fact that connects this feeling to the neurological processes she *can* recognize in her brain. For Chalmers, such difficulties underscore the *hard problem* of consciousness, that "there is nothing that we know more intimately than conscious experience, but there is nothing that is harder to explain."[67] In any case, Nagel, Jackson, and Chalmers illustrate that we cannot simply take for granted that neurological information is the equivalent of a particular psychical experience, and both "Mary's Room" and the "What is it like . . . ?" argument have retained prominent—if contested—places in the philosophy of mind.

These arguments raise significant questions for media physicalism, the answers to which have important implications for the debate over physicalism more generally. If there are qualia for which neurological research cannot account, then what might be left out of MRI studies of media use? If there is *something there is like* to be a person reading a book, then to what extent can we say that a scan of that person's brain gives us some—or any—picture of that experience? What is it that is being scanned? What is it that is being experienced? What does it mean to talk about "information overload" on the level of brain states, and how is this similar to or different from talking about a human experience of information overload? Even for those, like Christian Ruckmick, who believe that people are not the best judges of their own experiences, what does it mean to identify emotional stimulations in a person's physiology if that person does not herself identify or personally *feel* those supposed emotions?

A specifically media take on Jackson's thought experiment helps to illustrate the importance of these questions. What if Mary's room, rather than being free from color, is completely without electronic media, including television? Rather, she learns about neuroscience and physics from papers passed through a hole in her wall. By the time she exits her room, she is an expert on the chemical, physical, and biological changes that occur in the brain while watching television, as well as on the precise physical properties of television itself. When she sees television for the first time, does she learn something new?

This should presumably be enough to satisfy the thought experiment; however, anyone who knows a little about television should see that the question is not this simple—that "television," like Carr's "books," is not so coherent an object. What television does she watch? If it is 1950s American television, with three commercial networks broadcasting mostly live versions of variety shows and situation comedies, she is likely to understand her experience differently than if she sees a whole range of channels showing different "reality" television programs. The television in still different time periods and different countries would likely give her a still greater range of experiences. If by some strange quirk of fate, the first television she sees is broadcasting lectures on neuroscience, she might think to herself, "Television is an educational apparatus for teaching at a distance." While there would be some truth to this observation, she would likely change this appraisal as she saw more and different television programming. "Television is a vast wasteland"; "Television is high culture for the masses"; "Television is a tool of government propaganda"; and "Television is part of the fourth estate" are all possible observations she could make, to greater or lesser degrees, and to be modified or amplified with further and different kinds of viewing. In essence, she would *learn* something with each new viewing, until she gradually felt more and more confident that her knowledge of television approximated that of the people around her—though their experience would be changing with Mary's in relationship to ongoing changes in television's programs and institutions.

This raises the question of what precisely it means for Mary to *experience television*. If the first thing Mary sees is the chocolate factory episode of *I Love Lucy*, but she knows nothing about Lucille Ball or situation comedies in general, has she experienced television in the same way as other American television viewers of her age group? What if Mary grows up on a space station with a television set that broadcasts information and homemade entertainment programs, all produced by people who have had no contact with Earth, only to land in America, where she knows nothing not only about the history of television programming, but about the culture, country, and planet more widely? With absolutely no context for the programs she sees, including their place in a larger history of storytelling that stretches back at least as far as Aristotle, what precisely does her television *watching* entail? What does she learn? To return to our original, television-deprived Mary, what if instead of learning about the neuroscience of television viewing and the physics of television itself, she learns all about the history of television programming? That is, what if her wall hole has been filled with *TV Guides* and weekly and even daily descriptions of contemporary television programs (perhaps even whole scripts) for as long as she can remember? When she sees television for the first time—now as a virtual insider—how does she experience it? Does she learn something new, and if so, what?

From the standpoint of Nagel and Jackson, all these Marys would learn something—experiencing some unique set of qualia—when they see television. However, I would argue that the final Mary, the avid reader of *TV Guide*, would in a certain sense learn the least. While she would never have seen a television, because she is fully immersed in television culture, she would have a rich sense of its stories and discourses and would in an important way already be "experiencing" television viewership. She could converse about television and about the many larger cultural concepts on which it touches with a particular level of familiarity. Her reaction would likely be limited to something to the extent of "Wow, that's what Lucy looks like?" Space Station Mary would likely learn significantly more from her first experiences with American television, although the qualia associated with 525 lines scanned at 30 frames per second would not be among her lessons (her space station has an analog television). Neuroscientist Mary—deprived of television, as well as, let's assume, *TV Guide*—would learn still more. Her knowledge of brain physiology would tell her nothing about the specific programs she would see and—if we take Nagel, Jackson, and

Chalmers over Dennett and Churchland—little more about her overall feelings of experiencing them. Both the culture and qualia of the television she experienced would be new to her.

If *TV Guide* Mary indeed learns the least, then her situation suggests the importance of historical context to the experience of television viewing. Television makes sense in a vast web of cultural, linguistic, and textual references. On the one hand, television programs constantly reference other television programs, borrowing aesthetic and narrative conventions, economic structures (e.g., advertising), and so forth. At the same time, they draw on extant cultural values, aesthetic conventions, behaviors, codes of dress, and other general fashions of the time. For these reasons, older programs often look odd to later viewers, who see the cultural and historical situatedness of these earlier programs even as they view their own period's programs simply as natural representations of the world. If we ignore this historical context—or try to control for it, as much media physicalism will—we leave out a substantial way that everyday people experience television.

If this can be said of the experience of technology, it can be said of the experience of emotion as well. What if Neuroscientist Mary learns all about the brain physiology of *love*, but, being alone in her room, never herself experiences it? When she leaves her room and feels love for the first time, does she learn something new? As with the television example, if the questioning stops there, it misses much of what is most important about Mary's experience. When and with whom does Mary fall in love? If this is the eighteenth century, she might well fall in love with someone who writes love letters about her and addresses them to her sister. At the beginning of the twenty-first century, the same letter writer will likely strike Mary as creepy. What if Mary finds herself feeling love for another woman? Does she feel that she was just born that way? Does she feel that she's made a choice? Does she feel ashamed? Does she feel that she has a psychiatric illness that can be cured? Much of this will depend on the moment in which she lives and the messages that surround her—on television and elsewhere—whether within or beyond the walls of her room.

This historical context is not simply an additional *quale* (the singular of qualia) of technological or emotional experience. It is a larger framework in which various qualia are felt, interpreted, and made sense of. For the idea that our culture is "speeding up" to have any valence in our everyday lives, that experience needs to happen against a background in which things are assumed to have been slower in the past. The *feeling* of information overload, therefore, is necessarily historical, at least for individuals who feel they can remember some previously slower moment in their lives. But these feelings have larger contexts as well, tied to messages that address the uses, importance, and larger cultural effects of various media. What if No-TV Mary's neuroscience lessons teach her the merits of media physicalism, offering her evidence of both the sublime power of television and the necessity for physiological research, via brain imaging, to understand its impact on people's emotions? When she emerges from her room and sees television for the first time, will she trust her own experience of it? Or will she assume that she needs to get an MRI so she can see just how the television is affecting her? Whether she thinks she learns something when watching television will be heavily influenced by her previously held beliefs about emotion and technology themselves.

None of this should suggest that there is one, unified cultural experience that everyone has when they use a media technology. Indeed, groups that are actively excluded in discussions of certain technologies—as was the case for women and radio in the 1920s—may find themselves struggling against the dominant culture in a whole range of ways. If Mary emerged from her room in the 1920s and, knowing nothing about the culture of radio at the time, assumed that she could become a radio announcer as easily as any man, she would quickly find a number of cultural blockades and assumptions in her way. Of course, these are just the sorts of challenges that media physicalism tends to ignore by focusing on the neural pathways drawn by this or that media technology.

If the qualia of an individual experience cannot be reduced to a set of brain states, then certainly this collection of social contexts cannot either. Where would the social history of television—episodes of *I Love Lucy*, Aristotelian notions of storytelling, or a general knowledge of "reality television"—live in the brain? What about the rhetoric of the technological sublime, with its often contradictory understandings of the various new technologies of the moment? What about the idea that women are not good radio announcers? We do not need to postulate a "shopping instinct" to explain how and why people are drawn to new technologies. Marketers, academics, and popular writers do fine jobs of explaining each technology's supposed benefits for the culture at large, as the previous chapters have shown. Likewise, various economic conditions may make certain technologies necessary, regardless of whether or not someone feels addicted to them. The more information was published in books, the more important reading became (and the more the rhetoric of the book, which Carr so clearly espouses, could begin to take shape). These larger contexts are external from the specific brain states of people, although their implications may influence, or be visible in, various neurological changes.

Taking these issues seriously should complicate not only the physicalist position, but Nagel and Jackson's phenomenological response to it. If we think of experiences as having a historical component, then we have to acknowledge that even "seeing red" takes place in a particular context. Assuming her room is in twenty-first-century America, to what extent does Mary *see red* if she has none of the context—stop lights, the American flag, Valentine's Day, communism, anger, fire trucks—in which the color makes sense to the world outside? From this perspective, one of Dennett's objections to the Mary thought experiment seems correct; he asks, "Are we really so sure that what it is like to see red or blue can't be conveyed to one who has never seen colors in a few million or billion words?"[68] I would argue that a person who had engaged in this level of conversation *could* fairly be said to have knowledge of color, provided—and here I argue contrary to Dennett—that this was not simply a protracted conversation about brain states and neural activities. Helen Keller, who saw color only briefly as a child before she was stricken with blindness, said that she could still find pleasure in the beauty and colors of paintings. "I have at least the satisfaction of seeing them through the eyes of my friends, which is a real pleasure," she writes in her autobiography.[69] In an important way, Keller saw red as fully, if not more fully, than a Mary (neuroscientist or not) who has been deprived of this context. Of course, this "vision" would include the cultural baggage that surrounded it. Even if someone in the 1920s could not hear a voice on the radio, if he or she stayed up on popular or academic writing on the subject, that person would likely still *know* that men sounded better than women.

In terms of the questions that animate this book, taking this larger context into consideration means theorizing media in a way that foregrounds the historical and rhetorical situatedness of our experiences with technology and emotion. Carr, Johnson, and Small and Vorgan—like Christian Ruckmick, Carl Seashore, and the Keystone View Company before them—are both producers and products of particular rhetorics of technology and emotion. Understanding how their messages fit within and frame a set of wider cultural attitudes means taking a rhetorical approach to media and emotion themselves. Doing so should offer a more self-conscious understanding of our own discussions and illustrate not only how these messages coalesce in the first place, but what kinds of practices, experiences, ways of thinking, and groups of people are left out in the process.

Physicalist Problems and Consequences

It is difficult if not impossible to "prove" a philosophical thought experiment. The above examples get at fundamental assumptions about the nature of mind that are extremely hard to resolve. People committed to a media physicalist perspective are likely to find that neuroscience television Mary learns the least from seeing television for the first time, just as she could presumably tell us whether a chicken likes Pink Floyd (or even a person—it might be that you like Pink Floyd but don't realize it!). Although I doubt that I can provide counterevidence that would prove this position untenable for those who hold it as dearly as Ruckmick, Carr, Johnson, or Dennett likely would, I want to suggest three kinds of issues that plague media physicalism and that are demonstrated in various ways in the previous pages of this book. The first set of issues are *empirical*, the second are *ethical,* and the third *disciplinary.*

First, despite media physicalism's suggestions that physiological reactions are somehow the truth of media experience, there is strong anecdotal evidence that the cultural aspects of these media experiences are equal to, if not more important than, the technical ones. In his 1962 book *The Gutenberg Galaxy*, Marshall McLuhan cites a study by John Wilson that suggests that "preliterate" Africans are not capable of understanding the technical grammar of films.[70] Wilson had been working for the Gold Coast Department of Information when William Sellers, a health officer and co-head of the British Colonial Film Unit, had developed and shown a number of educational documentary films in sub-Saharan Africa. In the excerpt that McLuhan cites, Wilson recalled a famous incident involving an African audience watching a Sellers film about getting rid of standing water from a house. When asked to talk about the film, Wilson explained, the audience focused all of their attention on a chicken that had briefly run across the screen. From the perspective of Wilson and Sellers, these viewers had only seen a small aspect of the entire image they were shown.

Wilson offered that African viewers' attention to this minute detail had to do with their inability to take in the whole of the picture. As he explained in an excerpt quoted by McLuhan, these viewers "hadn't seen a whole frame—they had inspected the frame for details."[71] For McLuhan, Wilson's observations serve as evidence that "literacy gives people the power to focus a little way in front of an image so that we take in the whole image or picture at a glance." However, he continues, "non-literate people have no such acquired habit and do not look at objects in our way. Rather, they scan objects and images as we do the printed page, segment by segment. Thus they have no detached point of view. They are wholly *with* the object."[72] These examples provided McLuhan with evidence of the fundamental psychological differences between "oral" and "literate" cultures, though he was not alone in offering such arguments. This conception of African film viewership had found some powerful support in the 1940s and 1950s, as William Sellers created a widely followed version of simplified filmmaking that could presumably be comprehended by these less developed audiences.[73]

If it were true that preliterate Africans cannot understand film, that inability could provide support for media physicalists' claims about the essentially psycho-physiological dimensions of media use. The African brain—unaccustomed to film viewership—would simply be wired in a way that prevented it from seeing the multidimensional aspects of a film image. By the same token, brains in the West, and in other cultures with film, would have already been rewired by film viewership. However, as the historian James Burns demonstrates, Sellers's claims about African reactions to his films were often not as clearly supported as they appeared, and subsequent research—even in Sellers's own time—did not similarly demonstrate Africans' supposed film illiteracy. Critics of Sellers's position began to suggest that some of these responses—such as audiences' supposed concern that the people depicted on screen suffered from the "giant mosquitoes" depicted in a close-up—needed to be read ironically, perhaps even as jokes on the Sellers experiment itself. In fact, Burns explains, so many people began to criticize the Sellers film style as boring, pandering, and oversimplified, that it was eventually dropped completely.[74]

Even as Sellers was producing his films, the British anthropologist P. Morton-Williams offered a very different analysis of African film reception. "It seems quite evident that the physiological aspect of the problem can be ignored; that all audiences can see what is projected on to the screen, after a very short period," Morton-Williams wrote in a study that would help to establish the weaknesses of Sellers's position.[75] Later research supported Morton-Williams's position much more so than Sellers's. A study published in the *Journal of Communication* in 1988 explored how members of the Pokot tribe of Kenya with very limited experience of visual media responded to videotaped stories of familiar scenes. One group of viewers watched a video in which the camera remained relatively stationary; the other group watched a video in which "editing was used to make shifts in perspective and to enlarge details within the scene," containing "the frequent alteration of close-ups, medium shots, long shots, and zooms."[76] Ultimately, the study found little difference between the attention or memory of either group of viewers, suggesting that these first-time viewers were well equipped to deal with these more presumably complex editing techniques. That is, their brains were not somehow unable to comprehend the aesthetic conventions of film, despite their illiteracy and lack of experience with motion pictures.

Jenna Burrell's recent book, *Invisible Users: Youth in the Internet Cafés of Urban Ghana*, provides further evidence that the cultural aspects of media use may be more important than the more narrowly techno-physiological ones.[77] Burrell conducted an ethnography of nonelite, urban youth who frequented Internet cafés in Ghana's capital city, Accra. Her field research convinced her that "mastery over such a sociotechnical system for these youth turned out to be more than simply a matter of grasping technical features of the Internet and its interfaces."[78] Her book, in fact, suggests a wide range of more broadly cultural messages that help to materialize the Internet experiences of the young Ghanaians she interviews. Among other things, as these youth seek cross-cultural contact with other Internet users, they must negotiate the ways that their identities as Africans have already been constructed on a global—primarily Western—stage. Some of the youth

Burrell interviews were upset when other users ended their chats as soon as they revealed themselves as from Ghana, while others used these preconceived notions of Africa as a way of building "419 scams"—spam e-mail–based attempts to get people to wire them money.

Burrell found these youths' Internet use to be culturally circumscribed in still other ways as well. As she explains, "the delayed introduction of the Internet in Ghana meant that Ghanaian users entered into online environments that had, to some extent, already been staked out and normalized before their arrival."[79] These youth often found it difficult to understand the particular "social distance" expected by the cross-cultural users with whom they interacted, which was not always consonant with their own senses of communication. In fact, the specific details of their own media use—the fact that they were paying for their connection and had a limited amount of time that they could stay online—forced many of these youth to enter into relationships that, to the users on the other end of the line, seemed rushed. In a like way, Burrell demonstrates, "rumors" about the Internet that circulated among users in Ghana also played an important role in how they made sense of their own Internet usage. Stories of users who got rich through 419 scams had a powerful purchase on these youths' imaginations, even as their own experience did not generally bear these rumors out. Such examples offer powerful evidence of the role of various cultural forces in lending shape to the Internet experiences of the youth Burrell analyzes. Neither primitive minds without an ability to use Internet technology, nor twenty-first-century "netizens" who are immediately liberated by their computer use, these youth occupy a complex and contested cultural and physical space.

If Internet rumors and Western stereotypes about Africa can shape someone's media experience, then the more purely physiological accounts of media physicalism must certainly miss something. Would any amount of physiological knowledge allow neuroscientist Mary to understand these cultural forces' influence on Ghanaian youths' Internet use? If she were to emerge from her room with full knowledge of the Internet, wouldn't she also need to know precisely by whom the Internet had been built, and with what cultural values? By the same token, she would likely need to consider some specific information about *herself.* If she is Ghanaian, she might find her attempts to interact with other users quickly aborted. In any case, it's difficult to imagine how a brain scan, psycho-galvanometer, or other emotion-measuring device could tell us precisely how Mary—or any other Internet user—would make sense of these larger cultural matters.

The example of African cinema suggests a central ethical issue with media physicalism as well. As technological development becomes conflated with physiological development, media physicalism becomes supposed evidence for all sorts of arguments about modernity and civilization. As Burns demonstrates, Sellers and the various colonial filmmakers who followed his lead were more than willing to assume the technological backwardness of African audiences. This was particularly true for white theorists in South Africa, who supported Sellers's claims long after they had been discounted by much of the wider world. In championing Sellers's arguments about Africans' inability to view films, these theorists "were hinting at much more fundamental intellectual limitations inhibiting Africans from comprehending modern media. From this assertion, it was a short step to the conclusion that Africans were incapable of participating effectively in a technologically sophisticated, democratic society."[80]

As much of the previous examination demonstrates, such claims about racial, class, and gender differences are closely entwined with media physicalism more generally. The capacity of stereoscopes to cultivate the emotions of their users depended on a parallel rhetoric regarding the technological and emotional backwardness of people in China, India, Ireland, Italy, and elsewhere that stereoscope users could "travel." The supposedly friendly voice of the radio announcer and of the new public speaking in general was loaded throughout with assumptions about race, class, and gender. The movie excitement that most frightened Christian Ruckmick was evidenced by especially "exotic" examples of apparently foreign others. And such assumptions don't end with the early twentieth century, as the examples of Johnson, Carr, and especially Small and Vorgan demonstrate.

These examples are not somehow "misuses" of media physicalism. Rather, they express the cultural logic and larger rhetorical ecology that sustain the basic media physicalist premise. Discussing physicalism more broadly, the philosopher Naomi Scheman argues that its focus on the inner workings of single individuals "is deeply useful in the maintenance of capitalist and patriarchal society," which depends in many ways on isolating individuals—in theory, if not in practice—from the social structures in which they take place.[81] Similarly, in her analyses of techno-science, Donna Haraway demonstrates that a kind of biological reduc-tionism that she calls alternatively corporeal fetishism and gene fetish-ism disavows a wide range of institutions and practices that serve to continue various sorts of cultural power imbalances.[82] Approaches that reduce people to their physiological properties create universal subjects of a particular kind, abstracting them into similar collections of neuro-biological impulses. As Michael Warner has demonstrated, this kind of "self-abstraction" favors people who most clearly reflect the dominant culture.[83] While it may be easy for a white, middle-class American male to say that people's race, class, or gender don't matter when they go online, a Ghanaian youth's experience might tell him or her otherwise. When we ignore these experiences, we ignore some central factors in the disenfranchisement of those outside the dominant culture.

These problematic abstractions of emotional meaning and technology are not unique to journalistic discussions like those of Carr and Johnson or even to media research itself. Questions about media play an impor-tant behind-the-scenes role in the work of Antonio Damasio, a neuro-scientist who has written several well-known accounts of the physiology of emotion.[84] To Damasio's credit, much of his work has sought to chal-lenge the view that emotion is somehow opposed to reason by demon-strating that it is central to rational decision making. Still, his research reflects some of the uncritical abstraction that comes from divorcing mediated emotion from culture and history. For one study he discusses in his book, *Descartes' Error*, Damasio and his colleagues monitored the skin conductance of patients with frontal brain damage as they viewed a series of photographic images. Their goal was to understand how these patients evaluated photographs with "high emotional content." Accord-ing to Damasio, "normals" generate strong skin conductance "when we view scenes of horror or physical pain, or photographs of such scenes, or when we view sexually explicit images."[85] Damasio's patients would see a series of "banal" images, "but every now and then, randomly, a slide with a disturbing image would appear"[86] depicting "social disaster, mutilation, or nudity."[87] The fact that these patients recognized that the images quali-fied as "disturbing"—but showed no concomitant change in physiology— Damasio offers as proof that their injuries prevented them from feeling the emotions called forth by the images.

Like Christian Ruckmick had decades earlier, Damasio assumed that he could pick out emotionally "disturbing" images that would create a standard set of reactions for all "normals" (Damasio's use of the term "we" shows that he includes himself and his reader in this generalized category—presumably people with frontal brain damage don't read books, or at least not his). But can relative "disturbingness" really adhere so strongly in an image itself, transcending culture and time? Even the so-called normals of today are unlikely to find disturbing the film images used by Ruckmick in his studies of emotional stimulation of movies. Would images of social disaster, mutilation, and nudity be read in the same way at all times? What particular depictions of nudity are enough to satisfy the "high emotional content" that this study presumed to account for? Can it really be identified for all "normal" people?[88] In assuming so, Damasio sweeps away the emotional components of culture even as he depends on them for his definition of disturbing imagery. The implicit understanding of a normal emotional response to an image, for both Damasio and Ruckmick (who was likewise studying the electrical conductance of skin), is the taken-for-granted emotionology of white, middle-class culture.[89]

Despite the similarities between Ruckmick and Damasio—both of whom, in fact, were faculty at the University of Iowa—America's rhetoric of media physicalism has taken these assumptions about technology and emotion in some especially problematic directions. Here, the rhetoric of the technological sublime and continuing concerns about emotional control encompassed in American cool reach a potent nexus. New communication technologies—read inevitably as more powerful, stimulating, and so forth—are imagined to be writing on the body and brain in especially strong ways. In seeing mediated emotion in this way, media physicalism has little choice but to posit the evolutionary superiority of "technologically advanced" cultures—whose bodies and brains will always be more powerfully written-upon than those farther down the technological and emotional ladder. In the same way, it has been difficult for media physicalism to shed a variety of gender stereotypes, which themselves depend on essentialist claims about the biological differences between men and women and the supposedly more emotional nature of women.

Even as physicalism offers a range of both implicit and explicit claims about gender, race, and class, in focusing on the individual, physiological dimensions of mediated emotion, it overwhelmingly places these questions on hold, if not completely off-limits. If scholars claim to study the neurological effects of the Internet in isolation from cultural attitudes, it is because they choose to ignore the essentially white, Western cultural values on which it is based—those values that are experienced so keenly by the Ghanaian users in Burrell's study. Only a *hypothetical* Mary can have no race or class or exist in a cultural vacuum without any kinds of media rumors. By turning every media user into just such a hypothetical case, media physicalism both ignores and perpetuates some of the most central features of a culture's media experiences, as well as the inevitable power imbalances that accompany them.

Finally, the extent to which scholars—and others—choose to hold these cultural questions at bay matters in a disciplinary sense as well. As the above chapters seek to demonstrate, scholarly ideas about the new media of the early twentieth century did not spring objectively from thin air. Rather, they themselves were the products of a broader rhetorical ecology that included concerns about emotional control, a sense that the world was speeding up, and the idea that new technologies carried a powerful electrical and emotional force. As the rhetorical theorist Kenneth Burke explained, "all laboratory instruments of measurement and observation are invented by the symbol-using animal."[90] Media physicalism of the type practiced by Seashore, Ruckmick, and others was a concrete expression of a set of anxieties impacting the wider culture.

When the social sciences decided to put their faith in emotion-measuring machines, they turned a blind eye to a range of forces that were shaping their own practices. Taking for granted the power of new technologies to stimulate a body's emotions—themselves imagined primarily as technical forces—created a range of paradoxes and problems, not the least of which were the ethical problems related to race, class, and gender explored above. When researchers *used technologies to study technologies*, they created a complex feedback loop that resulted in a variety of problems and contradictions for their own capacity to understand the place of media in their own work, let alone the wider culture.

Even as thoughtful a media scholar as Marshall McLuhan could fall victim to these contradictions. McLuhan had a wide appreciation for the cultural aspects of media use, particularly as influenced by his colleague Harold Innis, a political economist who considered the broad cultural and material conditions of the media he studied.[91] However, as McLuhan's discussion of Africans' inability to see films demonstrates, elements of media physicalism ran through his work as well. McLuhan's ideas of "sense ratios" and "hot and cool" media demonstrate his sometimes subtle physicalism. Hot media were those that, primarily because of the high-fidelity nature of their image, required little participation on the part of users. The finely printed pages of post-Gutenberg books filled in much mental information that would have been absent at earlier moments in time. Cool media had lower-fidelity images that needed to be pieced together by users—such as the 525 scanned lines of analog television—and thus demanded higher participation.[92] By framing "participation" in terms of the internal mental activities of basic information processing, McLuhan laid the groundwork for the kinds of neurological claims made by Carr and Johnson, who use his ideas to support their seemingly different arguments about the benefits and drawbacks of technology.

As McLuhan became increasingly interested in sensory processes, he put more and more emphasis on scientific and neurological research. In *Laws of Media*, published in 1988—eight years after McLuhan's death—Marshall and his son and coauthor, Eric McLuhan, connected the media effects they noted to the different hemispheres of the brain. Drawing on neurological research that characterized the brain's right hemisphere as "simultaneous, holistic, and synthetic," while associating the left hemisphere with "linearity and sequentiality," the McLuhans identified the right hemisphere as the "acoustic (qualitative)" side of the brain, and the left as the "visual (quantitative)" side.[93] With this theory in hand, the effects that McLuhan had identified with oral and literate culture became explicitly neurological. By emphasizing visuality and linearity, the printing press actively engaged the left hemisphere of the brain. Print-era people assumed a homogenized scientific perspective that robbed them of the synthetic views of oral cultures, the McLuhans' argument went, because their brains had been rewired by their technological use.

In championing this neurological research even as they critiqued the perspective of linear, scientific thinking, the McLuhans fell prey to some of the same contradictions at work in the thinking of Ruckmick and Carr. This is evident in the McLuhans' discussion of the Shannon-Weaver model of communication, a later version of the transmission model that had dominated thought on the media when Ruckmick undertook his research.[94] Associating this model with the limitations of left-hemisphere thinking, the McLuhans write that Shannon-Weaver "is a kind of pipeline model of a hardware container for software content. It stresses the idea of 'inside' and 'outside' and assumes that communication is a kind of literal *matching* rather than resonant *making*."[95] As I demonstrate in the previous chapter, however, there were close connections between this kind of linear thinking about communication and the appeals to physiology made with such technologies as the psycho-galvanometer. In failing to theorize the brain scan as its own kind of technology—closely tied to the thinking of Shannon-Weaver—the McLuhans missed an important way in which their neurological approach supported the very linear perspective they claimed to be critiquing.

Reflecting other ideas discussed in the previous chapters, in his recent book *Perplexities of Consciousness*, Eric Schwitzgebel offers a sustained attempt to demonstrate the unreliability of introspection (despite—as he admits—the method's defeat in the early twentieth century). As an example of people's inability to understand their conscious experience, he offers evidence of competing studies about whether people dream in color or black and white. At different periods, he clearly illustrates, cultural consensuses have changed about which is the case. As a result, he argues, "to determine the coloration or non-coloration of the dream-world proves surprisingly difficult—pending at least, substantially more sophisticated psychological or neuroscientific research."[96] In another example, Schwitzgebel offers a criticism of introspection quite in line with the claims made by Christian Ruckmick, suggesting that people cannot understand their inner lives because their emotions get in the way of their introspective judgments. "If emotionality is enough consistently to undermine the reliability of our judgments about emotional experience," he claims, then introspection is unreliable.[97]

The ideas that introspection can be false—especially as compared to some neurological "truth"—and that emotions impair people's judgments are the very sorts of cultural assumptions that influenced the often problematic notions of media physicalism in the early twentieth century. In the philosophy of mind, everyday people's sense of their own experience is referred to as "folk psychology"—and it is one of the positions against which physicalism positions itself. Physicalism seeks to replace a supposedly flawed or incomplete folk psychology of the mind with a more scientifically accurate one. Jerome Bruner, an early champion of the cognitive revolution in psychology who would later object to its overly scientific emphasis, has been extremely critical of this notion of folk psychology. As he writes, "antisentimentalist fury about folk psychology simply misses the point. The idea of jettisoning it in the interest of getting rid of mental states in our everyday explanations of human behavior is tantamount to throwing away the very phenomena that psychology needs to explain."[98] Rather than simply dismissing folk psychology, Bruner asks psychologists to focus on how it is structured by various cultural *narratives* and the kinds of problems and possibilities it creates for people's lives.

In the terms of my analysis, and similarly to media physicalism, folk psychology could be understood as a kind of rhetorical ecology—a collection of stories that tell people how to make sense of their own and others' mental lives. As I illustrate, however, media physicalism is hardly confined to "everyday people"—if they believe in it at all. Still, its particular telling of stories about media and emotion can have a powerful effect, not the least of which is on scholars and theorists themselves. Without the self-reflection that characterizes much of the rest of his book, Schwitzgebel never considers how his own conception of emotion—as a kind of bodily contamination—might be influencing how he sees his subject (and himself). Likewise, he never asks how *cultural conversations* about color in film and television—both of which he addresses as variables in his study—might themselves be playing a role in people's perceptions of their dreams, focusing instead on a static conception of media and a "true or false" idea of human consciousness.

If scholars do not reflect upon the cultural questions that surround the media they explore, they risk blindly repeating the assumptions at work within their culture, or in the more narrow academic circles in which they take part—themselves part of the larger culture. If culture matters for any of the Marys I discuss above, than it must matter for scholars as well.

Physicalist Policy

The rhetorical power of media physicalism comes not only from the ways its assumptions are repeated by academics, journalists, clergy, and others; it also comes from its concrete realization in a range of practices and policies. Radio announcer examinations made it extremely difficult for most people to earn a spot on air. Especially when vocal frequencies were taken into account, certain voices would simply fail a basic technical test and be deemed unfit for broadcast.

In a like way, the Telecommunications Act of 1996, the legislation that replaced the Communications Act of 1934, repeated much of the earlier act's vocabulary, as well as its assumptions about the form and meaning of communication technology. If the 1934 act legislated a transmission model of communication, the 1996 act did so even more wholeheartedly, as can be seen in the definition of the act's principal term: "The term 'telecommunications' means the transmission, between or among points specified by the user, of information of the user's choosing, without change in the form or content of the information as sent and received."[99] This same point-to-point transmission of information is precisely how media physicalism imagines the communication process. Like media physicalism, by focusing on message transmission at the expense of larger social and cultural contexts, the 1996 act ignored and then perpetuated a whole range of power imbalances within the contemporary media environment. As Robert McChesney has demonstrated, the deregulatory nature of the 1996 act led to an unprecedented corporate concentration of communication technologies, ensuring that fewer voices would be heard in the broadcast media in particular.[100] As the 1934 act had also illustrated, when communication is seen merely as an act of transmission, it is easy to assume that the best role for the government, not to mention the public, is simply to step aside and let the transmissions proceed.

As close as the 1996 act got to protecting the public was section 5, the Communications Decency Act. In explaining and defending the act, which set up regulations for certain kinds of obscene and indecent material online, Jim Exon, one of its original authors, explained that "a child can get on the information superhighway and freely ride to on-line 'red light' districts that contain some of the most perverse and depraved pornographic material available."[101] Although portions of the Communications Decency Act were ultimately defeated in court, because its concern about children and pornography is the only place in the larger act that directly concerns emotion, it defines emotion for the Telecommunications Act of 1996 as a whole, and it does so through an essentially physicalist perspective. As with Carr, Johnson, Ruckmick, and Small and Vorgan, for this act of legislation, the emotional aspects of media use were a matter of direct harm or benefit to an individual's psychology (an approach that has seemed to define the early twenty-first-century FCC as well, which has shown much more concern for "obscenity" than for other questions about the cultural and emotional importance of communications messages). This perspective on emotion no doubt aided the larger power imbalances created by the 1996 act more generally, as the legislators never seemed to ask how opening up *access* to the media might itself serve the public's emotional well-being.

For all these reasons, the kind of rhetorical approach undertaken in this book is not only a way of analyzing talk about communication technology and emotion, but also a broader framework for understanding the cultural experience of both. When people engage with technology or experience an emotion, they take part in a thoroughly rhetorical process. As Mary's TV room example should suggest, when people watch television they negotiate a range of cultural discourses through which the technology and its content make sense. The same thing is true for emotions. How people respond to their own bodily sensations, the extent to which they express or suppress their emotional responses, and the ways they evaluate their emotional life more generally will be a complex process of identification between individuals and their cultures. Whether or not and how Mary experiences love will owe much to the historical period in which she lives—and the extent to which that love reflects the dominant values of her culture.

This should not suggest that people are simply cultural dupes buying into whatever ideas about technology and emotion circulate in the dominant culture. As this and the previous chapters have illustrated, these dominant values are often contradictory and incomplete, decrying certain elements of emotion or technology at the same time as they are celebrated. Still, as ideas solidify—even in their contradictions—they can form a range of important cultural norms with very strong material effects. A variety of activists in the early twenty-first century have been working tirelessly, but with fairly little success, to challenge the economic and regulatory effects of the Telecommunications Act of 1996. Today's ways of talking are tomorrow's legislative acts—often before the general culture has had a chance to reflect upon the significance of its vocabularies. A rhetorical approach to technology and emotion should make this reflection a central priority, calling attention to the possibilities and constraints at work in different kinds of discourses, including those whom these vocabularies disadvantage or disregard.

Watching television, surfing the web, reading a book, and various other media interactions are not simply questions of neurological chemistry or sensory experience more generally. Rather, they take place in dense historical contexts through which the experience makes sense. Like Johnson, Carr, and Small and Vorgan themselves, people experience media technology within a particular context for understanding technology itself. We can never know what it would be like to be a bat watching television, not only for the reasons Thomas Nagel suggests, but because there is no context for bat viewing that runs parallel to our own. Because bats do not have television, they presumably do not have ideas about television, let alone a whole history of talking about its cultural effects. The fact that we do is an integral part of our experience that cannot be controlled for or ignored.

Conclusion

In February 2011, IBM unveiled Watson, a computer system capable of answering questions posed in natural language, in three television episodes of the quiz show *Jeopardy!* Watson competed against two of the program's most successful contestants, Brad Rutter and Ken Jennings. As part of the opening broadcast, IBM included a short video about Watson's abilities and design. In discussing the computer's "stage presence," a voice-over explains the ideas behind Watson's avatar—a black screen depicting Watson's "face" as it answers questions:

> Based on IBM's smarter planet icon, the threads and thought rays that make up the avatar change colors and speed depending on what's happening during the game. When Watson feels confident in an answer, the rays in the avatar turn green. They turn orange when Watson gets an answer wrong. You'll see the avatar speed up and activate when Watson's algorithms are working hard to answer a clue. *It's the equivalent of watching a computer sweat.*[1]

In addressing why Watson might hesitate to answer a question when it is not confident in its response, Dr. David Ferrucci, the chief IBM scientist behind the project, suggested that Watson "doesn't want to look stupid"[2] (although this supposed fear did not stop Watson from answering "Toronto" to a question under the category of "U.S. cities"). From the standpoint of IBM, Watson was not only a highly developed thinking machine; it was a *feeling* machine.

If Watson has emotions, it must have felt happy about the results. It won the three-day event, and a million-dollar prize, by answering questions that amounted to $77,147, more than triple the scores of Jennings ($24,000) and Rutter ($21,600). Watson's success brought a flurry of debate about artificial intelligence and the place of computers in people's everyday lives. The *New York Times* columnist John Markoff worried that the power of the Watson system, and IBM's expressed desire to commercialize it, made it possible "to envision systems that replace not only human experts, but hundreds of thousands of well-paying jobs throughout the economy and around the globe."[3] In an article for the *Wall Street Journal*, the philosopher John Searle dusted off his classic "Chinese room" argument to refute the idea that what Watson did was anything like human thinking. According to Searle, Watson was manipulating symbols rather than engaging meanings. In the end, Searle argued, "Watson doesn't know it won on *Jeopardy*."[4]

Watson elicited some enormously celebratory comments as well. Among the most frequently cited positive implications of Watson's computing power was its ability to help humans deal with the excess information of the digital age. In a video segment from the second day of Watson's *Jeopardy!* appearance, the IBM project manager David Shepler explained, "There's so much content out there. Information overload is really the problem of our day." Jon Iwata, an IBM senior vice president, elaborated that Watson's programming is "about finding the needle in the haystack—the key insight that's useful."[5] Popular commentators celebrated Watson's information processing abilities as well, referring to it as a "Google-killer"[6] or describing it as "like Google with a little mind reading thrown in."[7] But for these celebrants, the Watsons of the future would be more than simply better search engines. As one *New York Post* writer explained, in the future, people in the medical industry would "have sick people spill their problems to the computer, talking exactly like they would to a human medical professional."[8] The computer would then comb through the Internet, online journals, and databases, taking advantage of a vast pool of medical knowledge to arrive at a presumably better diagnosis. For these writers, Watson seemed to offer that elusive technological solution to the information overload that had worried Walter Lippmann almost a century earlier. Watson would bring real insights out of an otherwise overwhelming mass of data.

In the context of the previous chapters, these conflicted commentaries on Watson evidence the complex contradictions of the rhetoric of media physicalism. The telegraph was empowering and frightening for one and the same technological power—its ability to transmit messages across vast distances and at great speeds. The same power that made stereoscopes, phonograph records, and motion pictures a threat to people's emotions made them the ideal tools for research on the media's emotional effects. With a similar sense of paradox, even as IBM and others celebrated Watson's "confidence," they also championed its lack of emotion. As the second-place finisher, Ken Jennings, explained in a commentary for *Slate*, "Watson cannot be intimidated. It never gets cocky or discouraged. It plays its game coldly, implacably, always offering a perfectly timed buzz when it's confident about an answer."[9] Presumably the perfect mix of confidence and emotional detachment, Watson was the ideal scientist that the psycho-galvanometer–wielding Ruckmick and Dysinger had aspired to be.

Does the fact that this same rhetoric has followed each of these new technologies suggest that each is no different from the others? Is Watson no more technologically advanced than the telegraph? Is the MRI just a fancy psycho-galvanometer? Is there no way to talk about the positive or negative aspects of technology without falling into the rhetoric of media physicalism? Do different technologies not have different impacts on people's emotions? Isn't it sometimes good to control one's emotions, as Watson does? Aren't some kinds of emotional expressions better than others?

It would be silly to pretend that there are no differences between various technologies. The fact that I am writing this book on a computer, which allows me to edit as I go, "find and replace" words or phrases, and "cut and paste" in ways that would be difficult or impossible with a typewriter, no doubt has had an effect on the way that I offer my arguments. The fact that television and radio allow the broadcasting of live events is a significant feature of their technological and social importance, as is the kind of mobile communication allowed by cellular telephones. If we lump all technologies together into one category, we miss both the unique potential of these different technologies and the different ways that each is received within a given culture or moment in time. That said, when we view technologies in isolation, as commentators often do when discussing the new technologies of their moment, we tend to uncritically reproduce the rhetoric of media physicalism, turning a blind eye to the ongoing history in which we make sense of our technological lives.

This larger history notwithstanding, there are some very good reasons why people are both hopeful and fearful about new technologies. In mechanizing the process of memory, writing allowed for new ways of thinking about a people's history. As Socrates's comments suggest, writing made it possible for the dead to speak to the living in a way that was not possible before. The telegraph sped up the process of sending these messages, and the computer increased the speed with which people could retrieve messages from a larger pool of assorted information. But what does it mean for a culture more generally to be "speeding up," especially when a nineteenth-century observer of the telegraph, an early twentieth-century observer of radio, and an early twenty-first-century observer of the Internet can all assume some kind of new *climax* of technological and cultural speed? We need to have a wider scholarly and cultural conversation about precisely what we mean when we talk about the increasing chaos, speed, or rate of interaction that gets consistently attributed to our new technologies. How much difference does it make if someone can get national news every morning—as was made possible by the telegraph—than if they can get it every few hours or minutes as now seems possible on cell phones or via the Internet? And how is this different from the day-to-day stress of someone without electronic media? What do we imagine to have increased, or not, and in what ways? Finally, in what ways might this sense of increasing speed be connected to other concerns or anxieties at work within our culture?

Our understandings of emotion need to have a similar sophistication. It is not enough to simply celebrate emotional control over emotional expressiveness, to do the reverse, or even to posit some middle ground of emotional balance akin to the friendly professionalism of the early twentieth-century radio announcer. We need to ask what kinds of emotional expressions are being championed or criticized, and for whom, at our own and other periods of time. A recent essay by Joshua Gunn demonstrates the emotional control still assumed to be a central value in women's public speech. He notes the journalistic and other responses to the grunting of female tennis players, observing that "female public grunting is received as sexual, not necessarily in the orgasmic sense, but certainly in the sense that grunting represents an uncontrolled, mindless body given over to the libidinal. Involuntary speech is threatening, it is unwieldy, it represents an absence of consciousness, the body on autopilot."[10] As was the case with the arguments for emotional control during the early twentieth century, this response to female grunting masks a number of larger social concerns, especially about athletic, muscled, female bodies, particularly in the context of deep anxieties about female sexuality. Noting such examples is not to say that no emotional speech should be seen as problematic or in need of control. Rather, we need to think about when, why, and how we deem such control necessary. Emotions can be good or bad, helpful or harmful, appropriate or inappropriate. Our beliefs about which is the case will often tell us more about ourselves—and our culture—than our actual emotional expressions will.

In the same way, simply because social scientists or their equipment may have helped to create some problematic conceptions of media and emotion does not mean that science is evil or that scientific analyses should be disregarded. Science offers an important perspective on media and emotion, as it does on a host of other topics. American medicine requires that doctors be scientifically informed on the physiology of the human body, and if I go to the hospital to have surgery on my heart, I will most likely be glad that this is the case. However, science is not the only or even the best perspective for every scenario. In subscribing to a predominantly scientific view, mainstream American medicine has pushed aside a host of other practices—holistic medicine, herbal therapy, acupuncture, religious healing practices, and so forth—that might be equally or even more helpful for certain people.[11] Likewise, if we have a predominantly scientific or technological view of emotion, we are likely to miss important aspects of our own and others' emotional lives that would help us think differently about the world in which we live.

In the context of American media studies, what I have referred to as "administrative research" likewise need not necessarily denote the decontextualized and often politically conservative approaches that I have addressed in the previous chapters. There is no inherent reason that research funded by the government or private corporations should focus on narrow questions of media effects that tend to turn a blind eye to problems of race, class, gender, and other such cultural issues. As I discuss in chapter 4, the Payne Fund motion picture research supported mutually by a private trust and a series of public universities originally contained highly politically charged arguments about the monopoly status of the Hollywood film industry. Peter Simonson and Gabriel Weimann have similarly demonstrated that media research in Paul Lazarsfeld's administratively oriented program at Columbia University could take a more "critical" orientation toward a range of problems.[12]

As the previous chapters illustrate, however, in both the early twentieth century and in our own time, there have been and are strong forces pushing American media research toward the narrower and more politically and economically safe questions pursued by such thinkers as Christian Ruckmick. The Payne Fund ultimately decided against critiquing the economics of Hollywood for fear of political reprisal. Scholars such as Glenn Merry and Smiley Blanton turned to technology-focused research as a means of demonstrating the legitimacy of their disciplines and research practices to both the academy and the world at large. As the rhetoric of the technological sublime has maintained a strong place in American culture, notions of scholarly truth and objectivity have taken a decidedly technological form. Here, establishing the legitimacy of a research project or claim increasingly means putting forward a set of technologically obtained, quantifiable data. Throughout the twentieth century, funding for the sciences grew exponentially in comparison to the humanities, where research on the larger cultural contexts of various phenomena tends to take place. By 1997, National Science Foundation grants were thirty-three times greater than those from the National Endowment for the Humanities (up from five times greater in 1979).[13] The beginning of the twenty-first century saw humanities researchers themselves increasingly embracing technological research and undertaking the kinds of experiments that Carl Seashore had pursued at the beginning of the previous century.[14] Again and again, America's tradition of administrative communication research— that research that wins the funding of government agencies and private corporations alike—has focused on using technologies to analyze themselves. And because technological research methods inevitably raise technological questions about the topics they are used to explore, such approaches too often reduce what should be complicated cultural processes to simplistic transmissions of one or another set of electrical impulses.

There are some real dangers to ignoring the larger cultural contexts that tend to be bypassed by technology-centered thinking about communication. Seeing both science and technology as a kind of endless process of advancement—whether this is seen as a positive moment toward a more enlightened, informed world, or a negative growth of information overload—casts those without advanced technologies as somehow primitive and backwards. The often troubling statements about race, class, and gender that I discuss in the previous chapters illustrate what happens when a narrative of technological advancement is accepted uncritically. When this narrative is paired with a physiological account of emotion, the problems become all the more pronounced—the brains and bodies of people with particular technologies can be presented as higher on the evolutionary scale than those without them. But does a teenager playing video games and talking on a cell phone really "multitask" more so than an auto mechanic, who needs to monitor a number of different electrical and mechanical systems in the process of working on a car (and did so even before computers were introduced into car design)? I can imagine a number of scenarios in which a hunter-gatherer would be a faster thinker and better at multitasking than I would. Narratives of technological advancement tend to oversimplify what should be some very complex questions, too often succumbing to—and perpetuating—hackneyed claims about gender, class, and racial differences.

The predominant American conceptions of media technology and emotion have had other consequences as well. As I discuss in chapter 1, during the nineteenth century, a number of people successfully sued telegraph companies over the emotional distress caused by undelivered messages. The courts that found on their behalves saw emotion as something highly public, shared, and social, in line with the Victorian conception of feelings that still had a prominent—if waning—place in American emotionology. Telegraph companies could be held responsible for their customers' emotional distress, because emotion was a broad social and public good. As shared emotion came more and more to be seen as something dangerous—the whirring of sentiments surrounding

Lippmann's "disenchanted man"—this public idea of emotion was replaced by the more private, individual conception that dominates in the early twentieth-century case studies I explore in the previous chapters. Likewise, while nineteenth-century American courts had a broad conception of the emotional and more broadly public responsibilities of telegraph companies, the American legal system gradually narrowed these duties to the technological transmission of information—reflecting the broader redefinition of technology and emotion that had taken place throughout the early twentieth century.

If the country had turned these nineteenth-century court decisions into legislation, the United States might have had a very different communication system. What if radio broadcasters, television producers, or Internet providers were told that they had a responsibility for the emotional well-being of the public? Would this change the kinds of content available to us? Would the same voices have access to the mainstream media, or would companies be forced to expand this as well? Instead of this model, the ideas put forward in the Communications Act of 1934, and reiterated in the Telecommunications Act of 1996, would largely support the narrowly physicalist ideas about technology and emotion developed in the early twentieth-century United States. American broadcasters' "public responsibilities" focused on how well they transmitted messages; when emotions were discussed, they primarily involved questions about "decency," rather than the broader issues implied by a more public conception of emotion. Of course, an American media system conceived under such a public idea of emotion would have its own problems as well. Still, if we want to be reflexive thinkers about media and emotion, their connections to each other and to the larger public need to be explicitly addressed, not only within media research, but within media messages, and within the media's economic and legal system.

In narrowing its attention to a technological and physiological conception of emotion, media physicalism ignores these larger legal, cultural, historical, and economic contexts. By focusing on, to borrow a phrase from Antonio Damasio, "the feeling of what happens"[15] to people *physiologically* when they engage communications media, media physicalism takes a *radically present-tense* stance that assumes away the historical contingencies that make particular media interactions possible in the first place. A hypothetical person, living in the now, and free from gender, racial, or class markings, can interact with a hypothetical medium, which itself has no apparent cultural or historical meanings. The previous chapters, however, suggest the importance of studying the feeling of what *doesn't* happen—tracing how various cultural developments close off certain kinds of emotional expressions or ways of thinking about technology or emotion. The fact that contemporary Americans do not typically send love letters to relatives of their love interests is at least as interesting as the neurological impulses that happen when they read something with "high emotional content." Attending to these lost possibilities should challenge us to reconsider the assumed normalcy of various emotional expressions, as well as to reflect on the extent to which we are happy with what our current stories ask us to believe about the possibilities for our emotional connections.

When we locate emotion exclusively, or even predominantly, in technologies or the physiology of the body, we largely abdicate responsibility for it. Schools that purchased stereoscopes, presumably buying into claims about the emotional and intellectual enlightenment they provided, allowed this technology to take the place of other educational practices. Today, educators often complain that students simply want to be entertained with high-tech gadgets, but schools at virtually every level continue to employ a range of devices imagined to make education more efficient, enlightening, or fun. By focusing on how motion pictures and other media technologies affected the physiology of specific individuals, the Payne Fund, as well as contemporary media physicalists, have pushed aside a range of larger philosophical and ethical questions about the media. If I score well on a psycho-galvanometer or MRI, or do not become violent after watching a violent movie, my stake in the issue is presumably over. The same perspective that suggests that media producers are responsible only for maintaining the technical qualities of their transmissions, suggests that I am responsible only for the physiological condition of my own body (which, Ruckmick and Dysinger suggest, I can never really know anyway). The stories told within our media and culture, and the ways they help make sense of our lives, become the responsibility of no one in particular. "Emotional control" becomes simply a euphemistic means of ignoring a host of larger social problems implicit in our understandings of emotion.

As I have suggested throughout, I believe that our discussions about how communication technologies advance our emotional lives have much to tell us about how we understand our world and ourselves. They also have material effects on our media system, communication policies, and larger culture. How we connect to each other and the potential problems and possibilities of these connections pose extremely powerful ethical, moral, and political questions. Although they are not always couched as such, our discussions about technology and emotion hit on some central issues about what we think it means to be human, to treat people well, and to live a good life. Ruckmick and Dysinger believed that they could use science to get beyond these questions about morality, but they were clearly advancing a particular ethic of emotional control and ideas about what kinds of movies were best for the public. I believe that our conversations will be more productive, useful, and critically reflexive when we make these political and ethical points more explicit. For me, this means tying our ideas to the larger climate in which they take place, noting who is empowered and disempowered by them, and doing our best to reformulate them to help those they disadvantage.

Friedrich Kittler has described a "discourse network" as "the network of technologies and institutions that allow a given culture to select, store, and process relevant data."[16] In introducing this concept, he hoped to move away from the sorts of discussions that he believed had dominated literary studies. Given the complex industrial and technological basis of books, he asserted, "what remain to be distinguished . . . are not emotional dispositions but systems."[17] The above chapters suggest that as scholars and as a wider culture, we need to be equally concerned with the *systems of emotional dispositions* that surround various communication technologies. What sorts of emotions are believed to be communicated by our various technologies, and what do these beliefs tell us about our assumptions about our own and others' feelings? As one of the newest technologies of our time, Watson—and the debates that surround it—gives us an important opportunity to reflect on our lives with technology and each other. Watson's performance on *Jeopardy!* demonstrated its great skill in answering questions. For us, it is all the more important that we *ask* questions—about our hopes and fears about technology, emotion, and ourselves.

NOTES

GUIDE TO ARCHIVAL SOURCES

CEMP: Charles E. Merriam Papers, Regenstein Library, Special Collections, University of Chicago.
DAUW: Division of Archives, University of Wisconsin, Madison.
ECSC: Emerson College Archives and Special Collections, Boston.
HDLP: Harold D. Lasswell Papers (MS 1043), Manuscripts and Archives, Yale University Library.
JALP: John A. Larson Papers, BANC MSS 78/160 cz, Bancroft Library, University of California, Berkeley.
JSSM: Johnson-Shaw Stereoscopic Museum, Meadville, PA.
PEBT: Papers of E.B. Titchener (14-23-545), Division of Rare and Manuscript Collections, Carl A. Kroch Library, Cornell University, Ithaca, NY.
PFIR: Payne Fund Inc., Records (MSS. No. 4315), Western Reserve Historical Society Manuscript Collections, Cleveland, OH.
UAUI: University Archives, Special Collections Department, University of Iowa Libraries, Iowa City.
WBCP: Walter Bradford Cannon Papers (H MS c40), Rare Books and Special Collections, Francis A. Countway Library of Medicine, Boston.

NOTES TO THE INTRODUCTION

1. Marvin Lee Minsky, *The Emotion Machine: Commonsense Thinking, Artificial Intelligence, and the Future of the Human Mind* (New York: Simon and Schuster, 2006), 7.
2. Jeremy N. Bailenson et al., "The Effect of Behavioral Realism and Form Realism of Real-Time Avatar Faces on Verbal Disclosure, Nonverbal Disclosure, Emotion Recognition, and Copresence in Dyadic Interaction," *Presence: Teleoperators and Virtual Environments* 15, no. 4 (2006): 368.

3. Byron Reeves and Clifford Ivar Nass, *The Media Equation: How People Treat Computers, Television, and New Media Like Real People and Places* (Stanford and New York: CSLI Publications and Cambridge University Press, 1996).

4. Joseph Gelmis, "The Film Director as Superstar: Stanley Kubrick," in *Stanley Kubrick: Interviews*, ed. Gene D. Phillips (Jackson: University Press of Mississippi, 2001), 95. On machine emotions in *2001*, see also Michael P. Nofz and Phil Vendy, "When Computers Say It with Feeling: Communication and Synthetic Emotions in Kubrick's *2001: A Space Odyssey*," *Journal of Communication Inquiry* 26, no. 1 (2002): 26–45.

5. Eric Alfred Havelock, *Preface to Plato* (Cambridge: Belknap, 1963); Walter J. Ong, *Orality and Literacy: The Technologizing of the Word* (London: Methuen, 1982); John Durham Peters, *Speaking into the Air: A History of the Idea of Communication* (Chicago: University of Chicago Press, 1999).

6. Jacques Derrida, "Plato's Pharmacy," in *Dissemination* (Chicago: University of Chicago Press, 1981), 62–172.

7. *Phaedrus* (230d). There has been some interesting debate over whether these thoughts and words about writing should be attributed to Socrates or to Plato. For a summary of these arguments, as well as a defense of seeing Socrates's claims about writing as distinct from Plato's position, see Dorrit Cohn, "Does Socrates Speak for Plato? Reflections on an Open Question," *New Literary History* 32 (2001): 485–500.

8. Deborah Lubken, "Remembering the Straw Man: The Travels and Adventures of Hypodermic," in *The History of Media and Communication Research: Contested Memories*, ed. David W. Park and Jefferson Pooley (New York: Peter Lang, 2008), 19–42.

9. *Phaedrus* (227b). These English translations are drawn from Plato, "The Phaedrus," in *The Collected Dialogues of Plato*, ed. Edith Hamilton and Huntington Cairns (Princeton: Princeton University Press, 1989).

10. *Phaedrus* (228e) (emphasis added).

11. Ibid. (263e).

12. Ibid. (275e).

13. Merritt Caldwell, *Manual of Elocution: Embracing Voice and Gesture Designed for Schools, Academies, and Colleges, as Well as for Private Learners* (Philadelphia: Sorin and Ball, 1846), ix.

14. Leo Marx, *The Machine in the Garden: Technology and the Pastoral Ideal in America* (New York: Oxford University Press, 1964).

15. Tom Lutz, *Crying: The Natural and Cultural History of Tears* (New York: Norton, 1999).

16. Peter Stearns and Carol Stearns, "Emotionology: Clarifying the History of Emotional Standards," *American Historical Review* 90, no. 4 (1985): 813. For a discussion of the strengths and weaknesses of the Stearnses' position, see Brent Malin, "Communication with Feeling: Emotion, Publicness, and Embodiment," *Quarterly Journal of Speech* 87 (2001): 216–35.

17. Steven R. López et al., "Cultural Variability in the Manifestation of Expressed Emotion," *Family Process* 48, no. 2 (2009): 179–94; Nalini Ambady and Jamshed Bharucha, "Culture and the Brain," *Current Directions in Psychological Science* 18, no. 6 (2009): 342–45; Hillary Anger Elfenbein and Nalini Ambady, "Universals and Cultural Differences in Recognizing Emotions," *Current Directions in Psychological Science* 12, no. 5 (2003): 159–64; Abigail A. Marsh, Hillary Anger Elfenbein, and Nalini Ambady, "Nonverbal 'Accents': Cultural Differences in Facial Expressions of Emotion," *Psychological Science* 14, no. 4 (2003): 373–76; Lisa Feldman Barrett, "Variety Is the Spice of Life: A Psychological Construction Approach to Understanding Variability in Emotion," *Cognition and Emotion* 23, no. 7 (2009): 1284–306; Hazel Rose Markus and Shinobu Kitayama, "Culture and the Self: Implications for Cognition, Emotion, and Motivation," *Psychological Review* 98, no. 2 (1991): 224–53.

18. For an exploration of issues surrounding women's expressions of emotion, see especially Alison M. Jaggar, "Love and Knowledge: Emotion in Feminist Epistemology," in *Gender/Body/Knowledge: Feminist Reconstructions of Being and Knowing*, ed. Alison M. Jaggar and Susan Bordo (New Brunswick: Rutgers University Press, 1989), 145–71.

19. Friedrich A. Kittler, *Gramophone, Film, Typewriter* (Stanford: Stanford University Press, 1999), 3.

20. According to Mumford, the approach to the world encouraged by magic helped pave the way for scientific thinking. He argues that "magic turned men's minds to the external world; it suggested the need of manipulating it; it helped create the tools of successfully achieving this, and it sharpened observation as to the results." Lewis Mumford, *Technics and Civilization* (New York: Harcourt, 1934), 40.

21. Stephen R. Fox, *The Mirror Makers: A History of American Advertising and Its Creators* (Urbana: University of Illinois Press, 1997), 39.

22. Thorstein Veblen, *The Theory of the Leisure Class: An Economic Study in the Evolution of Institutions* (New York: Macmillan, 1899).

23. On the larger history of advertising in America, see especially T. J. Jackson Lears, *Fables of Abundance: A Cultural History of Advertising in America* (New York: Basic, 1994); Roland Marchand, *Advertising the American Dream: Making Way for Modernity, 1920–1940* (Berkeley: University of California Press, 1985); and Fox, *The Mirror Makers*.

24. *Oxford English Dictionary*, 2nd ed.

25. Larry Tye, *The Father of Spin: Edward L. Bernays and the Birth of Public Relations* (New York: Crown, 1998); Edward L. Bernays, *Crystallizing Public Opinion* (New York: Boni and Liveright, 1923); Bernays, "Molding Public Opinion," *Annals of the American Academy of Political and Social Science* 179 (1935): 85.

26. Dale Carnegie, *How to Win Friends and Influence People* (New York: Simon and Schuster, 1936).

27. Ruth Vasey, *The World according to Hollywood, 1918–1939* (Madison: University of Wisconsin Press, 1997); John Trumpbour, *Selling Hollywood to the World: U.S.*

and European Struggles for Mastery of the Global Film Industry, 1920–1950 (Cambridge: Cambridge University Press, 2002).

28. Carl E. Seashore, "A Voice Tonoscope," *University of Iowa Studies in Psychology* 3 (1902): 18–28; Seashore, "The Tonoscope," *Psychological Monographs* 32, no. 3 (1916): 1–12; Seashore, "Phonophotography in the Measurement of the Expression of Emotion in Music and Speech," *Scientific Monthly* 24, no. 5 (1927): 463–71.

29. Paul Lazarsfeld, "Remarks on Administrative and Critical Research," *Studies in Philosophy and Social Science* 9, no. 1 (1941): 2–16.

30. Susan J. Douglas, *Inventing American Broadcasting, 1899–1922* (Baltimore: Johns Hopkins University Press, 1987); Douglas, *Listening In: Radio and the American Imagination from Amos 'n' Andy and Edward R. Murrow to Wolfman Jack and Howard Stern* (New York: Times, 1999); Robert McChesney, *Rich Media, Poor Democracy: Communication Politics in Dubious Times* (Urbana: University of Illinois Press, 1999); McChesney, *Telecommunications, Mass Media, and Democracy: The Battle for the Control of U.S. Broadcasting, 1928–1935* (New York: Oxford University Press, 1993); Paul Starr, *The Creation of the Media: Political Origins of Modern Communications* (New York: Basic, 2004); Erik Barnouw, *A Tower in Babel: A History of Broadcasting in the United States to 1933* (Oxford: Oxford University Press, 1966); Carolyn Marvin, *When Old Technologies Were New: Thinking about Electric Communication in the Late Nineteenth Century* (New York: Oxford University Press, 1988); Daniel J. Czitrom, *Media and the American Mind: From Morse to McLuhan* (Chapel Hill: University of North Carolina Press, 1982); James W. Carey, *Communication as Culture: Essays on Media and Society* (Boston: Unwin Hyman, 1989); Michele Hilmes, *Radio Voices: American Broadcasting, 1922–1952* (Minneapolis: University of Minnesota Press, 1997); Lisa Gitelman, *Always Already New: Media, History and the Data of Culture* (Cambridge: MIT Press, 2006); Gitelman, *Scripts, Grooves, and Writing Machines: Representing Technology in the Edison Era* (Stanford: Stanford University Press, 1999); Lisa Gitelman and Geoffrey B. Pingree, eds., *New Media, 1740–1915* (Cambridge: MIT Press, 2003); Friedrich A. Kittler, *Discourse Networks, 1800/1900* (Stanford: Stanford University Press, 1990); Kittler, *Gramophone, Film, Typewriter*; Jeffrey Sconce, *Haunted Media: Electronic Presence from Telegraphy to Television* (Durham: Duke University Press, 2000); Jonathan Sterne, *The Audible Past: Cultural Origins of Sound Reproduction* (Durham: Duke University Press, 2003).

31. David E. Nye, *American Technological Sublime* (Cambridge: MIT Press, 1994); Nye, *Electrifying America: Social Meanings of a New Technology, 1880–1940* (Cambridge: MIT Press, 1990); Wolfgang Schivelbusch, *The Railway Journey: The Industrialization of Time and Space in the 19th Century* (Berkeley: University of California Press, 1986); Marx, *The Machine in the Garden*; Wiebe E. Bijker, *Of Bicycles, Bakelites, and Bulbs: Toward a Theory of Sociotechnical Change* (Cambridge: MIT Press, 1995).

32. Jonathan Crary, *Techniques of the Observer: On Vision and Modernity in the Nineteenth Century* (Cambridge: MIT Press, 1990); Crary, *Suspensions of Perception: Attention, Spectacle, and Modern Culture* (Cambridge: MIT Press, 1999).

33. See, for instance, Patricia Ticineto Clough and Jean O'Malley Halley, eds., *The Affective Turn: Theorizing the Social* (Durham: Duke University Press, 2007); Melissa Gregg and Gregory J. Seigworth, eds., *The Affect Theory Reader* (Durham: Duke University Press, 2010); Sara Ahmed, *The Cultural Politics of Emotion* (New York: Routledge, 2004); and Daniel M. Gross, *The Secret History of Emotion: From Aristotle's Rhetoric to Modern Brain Science* (Chicago: University of Chicago Press, 2006).

34. Herbert Feigl, "Logical Analysis of the Psychophysical Problem: A Contribution of the New Positivism," *Philosophy of Science* 1, no. 4 (1934): 436; C. Wade Savage, "Obituary for Herbert Feigl," *Erkenntnis* 31, no. 1 (1989): v–ix. On the larger history of the Vienna Circle, see Viktor Kraft, *The Vienna Circle: The Origin of Neo-Positivism* (New York: Philosophical Library, 1953); and Friedrich Stadler, ed., *The Vienna Circle and Logical Empiricism: Re-Evaluation and Future Perspectives* (Dordrecht: Kluwer, 2003).

35. Theodor W. Adorno, "On Popular Music," *Studies in Philosophy and Social Science* 9, no. 1 (1941): 17–48; Max Horkheimer and Theodor W. Adorno, *Dialectic of Enlightenment* (New York: Herder and Herder, 1972).

36. Thomas S. Kuhn, *The Structure of Scientific Revolutions* (Chicago: University of Chicago Press, 1962). On the rhetoric of science more generally, see Leah Ceccarelli, *Shaping Science with Rhetoric: The Cases of Dobzhansky, Schrödinger, and Wilson* (Chicago: University of Chicago Press, 2001); Jeanne Fahnestock, *Rhetorical Figures in Science* (New York: Oxford University Press, 1999); Alan G. Gross, *The Rhetoric of Science* (Cambridge: Harvard University Press, 1990); and Alan G. Gross and William M. Keith, *Rhetorical Hermeneutics: Invention and Interpretation in the Age of Science* (Albany: State University of New York Press, 1997). For discussion of the rhetoric of inquiry as an area of study as well as examples of work in this vein, see John S. Nelson, Allan Megill, and Donald N. McCloskey, *The Rhetoric of the Human Sciences: Language and Argument in Scholarship and Public Affairs* (Madison: University of Wisconsin Press, 1987); John Lyne, "Rhetorics of Inquiry," *Quarterly Journal of Speech* 71, no. 1 (1985): 65–73; Herbert W. Simons, *Rhetoric in the Human Sciences* (London: Sage, 1989); Herbert W. Simons, ed., *The Rhetorical Turn: Invention and Persuasion in the Conduct of Inquiry* (Chicago: University of Chicago Press, 1990); John S. Nelson and Allan Megill, "Rhetoric of Inquiry: Projects and Prospects," *Quarterly Journal of Speech* 72, no. 1 (1986): 20–37; Donald N. McCloskey, *The Rhetoric of Economics* (Madison: University of Wisconsin Press, 1985); John S. Nelson, *Tropes of Politics: Science, Theory, Rhetoric, Action* (Madison: University of Wisconsin Press, 1998); and Milner S. Ball, *Lying Down Together: Law, Metaphor, and Theology* (Madison: University of Wisconsin Press, 1985). For a critique of the rhetoric of science and rhetoric of inquiry positions, see especially Dilip Parameshwar Gaonkar, "The Idea of Rhetoric in the Rhetoric of Science," *Southern Communication Journal* 58, no. 4 (1993): 258–95; and Gaonkar, "Rhetoric and Its Double: Reflections on the Rhetorical Turn in the Human Sciences," in

Simons, *The Rhetorical Turn*, 341–66. For responses to this critique, see Alan G. Gross, "What If We're Not Producing Knowledge? Critical Reflections on the Rhetorical Criticism of Science," *Southern Communication Journal* 58, no. 4 (1993): 301–5; Michael Leff, "The Idea of Rhetoric as Interpretive Practice: A Humanist's Response to Gaonkar," *Southern Communication Journal* 58, no. 4 (1993): 296–300; and Lawrence J. Prelli, "Rhetorical Perspective and the Limits of Critique," *Southern Communication Journal* 58, no. 4 (1993): 319–27.

37. Charles Bazerman, *The Languages of Edison's Light* (Cambridge: MIT Press, 1999), 3.

38. Jenny Edbauer, "Unframing Models of Public Distribution: From Rhetorical Situation to Rhetorical Ecologies," *Rhetoric Society Quarterly* 35, no. 4 (2005): 5–24; Nathaniel Rivers and Ryan Weber, "Ecological, Pedagogical, Public Rhetoric," *College Composition and Communication* 63, no. 2 (2011): 187–218.

39. On this debate, see Lloyd F. Bitzer, "The Rhetorical Situation," *Philosophy and Rhetoric* 1, no. 1 (1968): 1–14; Richard E. Vatz, "The Myth of the Rhetorical Situation," *Philosophy and Rhetoric* 6, no. 3 (1973): 154–61; and Barbara A. Biesecker, "Rethinking the Rhetorical Situation from within the Thematic of Différance," *Philosophy and Rhetoric* 22, no. 2 (1989): 110–30.

40. Edbauer, "Unframing Models of Public Distribution," 13.

41. Ibid., 18. While I appreciate Edbauer's decision to use viral metaphors to discuss the complex relationships among component pieces of a rhetorical ecology, I also note that doing so evokes a sense of bodily disease not unlike some of the discussions of mediated emotion I explore in this book. Perhaps there is an implicit physicalism at work in contemporary rhetorical theory as well.

42. Carey, *Communication as Culture*.

NOTES TO CHAPTER 1

1. Benjamin Franklin et al., *Memoirs of Benjamin Franklin*, vol. 2 (Philadelphia: M'Carty and Davis, 1834), 266.

2. Benjamin Franklin to Peter Collinson, July 28, 1747, in *The Papers of Benjamin Franklin*, ed. Leonard W. Labaree et al. (New Haven: Yale University Press, 1959).

3. Benjamin Franklin, "Lightning Rods for St. Paul's Cathedral (June 7, 1769)," in Labaree et al., *Papers of Benjamin Franklin*.

4. Peters, *Speaking into the Air*, 78 (see intro., n. 4).

5. Elizabeth Barnard, "The Atlantic Telegraph," in *Heart Offerings* (Chatfield, MN: Elizabeth Barnard, 1883), 87.

6. George F. Hoar, "Address of Senator George F. Hoar," *American Missionary* 35, no. 11 (1881): 372.

7. "Correspondence," *Philadelphia Medical Times*, April 7, 1883, 480.

8. "Intellectual Effects of Electricity," *Eclectic Magazine*, December 1889, 856.

9. "Not So Dark as It Seems," *Christian Union*, November 4, 1874, 350.

10. Martin F. Tupper, "A Rhyme for the Atlantic Telegraph," *Friend's Review*, January 24, 1857.

11. Marx, *The Machine in the Garden* (see intro., n. 14); Perry Miller, "The Responsibility of Mind in a Civilization of Machines," *American Scholar* 31, no. 1 (1961–62): 51–69; Miller, *The Life of the Mind in America, from the Revolution to the Civil War* (New York: Harcourt, 1965). For an especially well developed inquiry into the technological sublime in America, see Nye, *American Technological Sublime* (see intro., n. 31).

12. Marx, *The Machine in the Garden*, 3 (see intro., n. 14).

13. Ibid., 203.

14. Schivelbusch, *The Railway Journey*, 59–61 (see intro., n. 31).

15. Edmund Burke, *A Philosophical Enquiry into the Origin of Our Ideas of the Sublime and Beautiful* (Oxford: Oxford University Press, 1990), 67.

16. Immanuel Kant, *The Critique of Judgement*, trans. James Creed Meredith (Oxford: Clarendon, 1952), 91.

17. Burke, *Philosophical Enquiry*, 66.

18. Samuel Watson, *The Clock Struck Three, Being a Review of the Clock Struck One, and Reply to It: Part II, Showing the Harmony between Christiany, Science, and Spiritualism* (Chicago: Religio-Philosophical Publishing House, 1874), 22.

19. John Bovee Dods, *The Philosophy of Electrical Psychology: In a Course of Twelve Lectures* (New York: Fowlers and Wells, 1852), 137.

20. On the history of electricity in America, including various cultural reactions to it, see Carey, *Communication as Culture* (see intro., n. 30); and Nye, *Electrifying America* (see intro., n. 31).

21. Charles Briggs and Augustus Maverick, *The Story of the Telegraph, and the History of the Great Atlantic Cable* (New York: Rudd and Carleton, 1858), 13.

22. "Punch: The Bermondsey Horror," *Littell's Living Age*, November 24, 1849, 382.

23. Carey, *Communication as Culture*, 207 (see intro., n. 30).

24. Peters, *Speaking into the Air* (see intro., n. 4); Kittler, *Gramophone, Film, Typewriter* (see intro., n. 19).

25. Miller, *Life of the Mind in America*, 48.

26. Schivelbusch, *The Railway Journey* (see intro., n. 31).

27. On the history of the telegraph, see especially Richard John, *Network Nation: Inventing American Telecommunications* (Cambridge: Belknap, 2010). On the specific relationship between the telegraph and the railroad, see Schivelbusch, *The Railway Journey* (see intro., n. 31); and Carey, *Communication as Culture* (see intro., n. 30).

28. "A Dissertation upon the Passion of Anger," *American Magazine and Historical Chronicle*, May 1744, 375.

29. "On the Power of the Passions," *New York Magazine*, June 1796, 281.

30. Edmund Clarence Stedman and Ellen Mackay Hutchinson, eds., *A Library of American Literature: From the Earliest Settlement to the Present Time* (New York: Webster, 1889), 391.

31. Douglas L. Winiarski, "Jonathan Edwards, Enthusiast? Radical Revivalism and the Great Awakening in the Connecticut Valley," *Church History* 74, no. 4 (2005):

738. Winiarski has done a great service in publishing a recently discovered entry by Samuel Phillips Savage, in which Savage describes what he witnesses at one of Edwards's revival services. In so doing, Winiarski helps to illuminate Edwards's centrality to the revivalist spirit of his time. As Winiarski explains, "Edwards is best known for his reasoned defense of evangelical Calvinism, yet he was an equally powerful revivalist who hovered above contorting bodies and rapturous groans" (689). On the larger issue of the Great Awakening, see Joseph Conforti, "The Invention of the Great Awakening, 1795–1842," *Early American Literature* 26 (1991); James F. Cooper Jr., "Enthusiasts or Democrats? Separatism, Church Government, and the Great Awakening in Massachusetts," *New England Quarterly* 65, no. 2 (1992); Thomas S. Kidd, "Daniel Rogers' Egalitarian Great Awakening," *Journal of the Historical Society* 7, no. 1 (2007): 111–35; Frank Lambert, "'Pedlar in Divinity': George Whitefield and the Great Awakening, 1737–1745," *Journal of American History* 77, no. 3 (1990): 812–37; Lambert, "The First Great Awakening: Whose Interpretive Fiction?," *New England Quarterly* 68, no. 4 (1995); Lambert, "The Great Awakening as Artifact: George Whitefield and the Construction of Intercolonial Revival, 1739–1745," *Church History* 60, no. 2 (1991): 223–46; Susan O'Brien, "A Transatlantic Community of Saints: The Great Awakening and the First Evangelical Network, 1735–1755," *American Historical Review* 91, no. 4 (1986); John F. Slater, "The Sleepwalker and the Great Awakening: Brown's Edgar Huntly and Jonathan Edwards," *Papers on Language and Literature* 19 (1983): 199–217; Robert Lee Stuart, "Jonathan Edwards at Enfield: 'And Oh the Cheerfulness and Pleasantness . . . ,'" *American Literature* 48 (1976); Douglas L. Winiarski, "Souls Filled with Ravishing Transport: Heavenly Visions and the Radical Awakening in New England," *William and Mary Quarterly* 61, no. 1 (2004): 3–46; and Miller, *Life of the Mind in America*.

32. Jonathan Edwards, *Thoughts on the Religious Revival in New England* (New York: American Tract Society, 1740), 240–41.

33. Ibid., 244–52.

34. Nicole Eustace, "'The Cornerstone of a Copious Work': Love and Power in Eighteenth-Century Courtship," *Journal of Social History* 34, no. 3 (2001); Eustace, *Passion Is the Gale: Emotion, Power, and the Coming of the American Revolution* (Chapel Hill: University of North Carolina Press, 2008). On the history of eighteenth-century marriage and family, including the relatively public nature of a range of relational communications, see Nancy F. Cott, "Eighteenth-Century Family and Social Life Revealed in Massachusetts Divorce Records," *Journal of Social History* 10, no. 1 (1976): 20–43; Ellen K. Rothman, *Hands and Hearts: A History of Courtship in America* (New York: Basic, 1984); Rothman, "Sex and Self-Control: Middle-Class Courtship in America, 1770–1870," *Journal of Social History* 15, no. 3 (1982); Daniel Scott Smith, "Parental Power and Marriage Patterns: An Analysis of Historical Trends in Hingham, Massachusetts," *Journal of Marriage and Family* 35, no. 3 (1973): 419–28; Jan Lewis, "The Republican Wife: Virtue and Seduction in the Early Republic," *William and Mary Quarterly* 44, no. 4 (1987): 689–721; Robert J. Gough, "Close-Kin Marriage and Upper-Class Formation in Late-Eighteenth-Century

Philadelphia," *Journal of Family History* 14, no. 2 (1989): 119–36; and Daniel Blake Smith, "The Study of the Family in Early America: Trends, Problems, and Prospects," *William and Mary Quarterly* 39, no. 1 (1982): 4–28.

35. "A Tear," *Rural Magazine*, January 19, 1799, 4.

36. "The Tear," *Weekly Museum* 12, no. 46 (1800): 4.

37. On the history of crying in Western culture, see Lutz, *Crying* (see intro., n. 15). On the conflicted history of masculinity in the United States, see Michael S. Kimmel, *Manhood in America: A Cultural History* (New York: Free Press, 1996); Gail Bederman, *Manliness and Civilization: A Cultural History of Gender and Race in the United States, 1880–1917* (Chicago: University of Chicago Press, 1995); E. Anthony Rotundo, *American Manhood: Transformations in Masculinity from the Revolution to the Modern Era* (New York: Basic, 1993); Christopher E. Forth, *Masculinity in the Modern West: Gender, Civilization and the Body* (New York: Palgrave Macmillan, 2008); and Brenton J. Malin, *American Masculinity under Clinton: Popular Media and the Nineties "Crisis of Masculinity"* (New York: Peter Lang, 2005).

38. Michael Leff and Margaret Organ Procario, "Rhetorical Theory in Speech Communication," in *Speech Communication in the 20th Century*, ed. Thomas Benson (Carbondale: Southern Illinois University Press, 1985).

39. Wm. T. Ross, *Voice Culture and Elocution* (New York: Baker and Taylor, 1890), xiv.

40. J. W. Shoemaker, *Advanced Elocution: Designed as a Practical Treatise for Teachers and Students in Vocal Training, Articulation, Physical Culture, and Gesture* (Philadelphia: Penn Publishing, 1898), 3.

41. Albert Bacon, *A Manual of Gesture: Embracing a Complete System of Notation Together with Principles of Interpretation and Selections for Practice* (Chicago: Griggs, 1881), 40.

42. Ibid., 37.

43. John Walker, *Elements of Elocution: In Which the Principles of Reading and Speaking Are Investigated* (Boston: Mallory, 1810), 305.

44. Peter Stearns, *American Cool: Constructing a Twentieth-Century Emotional Style* (New York: New York University Press, 1994), 53.

45. "Temper," *Boston Weekly Magazine*, September 22, 1804, 189.

46. "To the Editor of the *Christian Observer*," *Christian Observer, Conducted by Members of the Established Church*, September 1804, 546.

47. Miss Bowdler, "On Sensibility," *Philadelphia Repository of Weekly Register*, May 14, 1803, 157.

48. A. L. M., "The Redemption of Labor," *Mechanic Apprentice*, December 1845, 59.

49. "Extract of a Letter from the Reverend Gideon Blackburn, Missionary to the Cherokees, to the Chairman of the Committee of Missions, Dated April 24, 1805," *Virginia Religious Magazine*, September 1, 1805, 316.

50. "Traits of Indian Character," *American Monthly Magazine and Critical Review*, December 1818, 157.

51. "Authentic Memoirs of the Conversion of a Negro," *United States Christian Magazine*, February 1, 1796, 117.

52. *So Relle v. Western Union Telegraph Company*, 55 Texas 308 (1881). On the larger history and legal implications of this case, see Brenton J. Malin, "Failed Transmissions and Broken Hearts: The Telegraph, Communications Law, and the Emotional Responsibility of New Technologies," *Media History* 17, no. 4 (2011): 331–44.

53. *Reese v. Western Union Telegraph Company*, 24 N.E. 163 (1890).

54. *Green v. Western Union Telegraph Company*, 49 S.E. 65 (1904).

55. *Western Union Telegraph Company v. Moore*, 76 Texas 66 (1890).

56. *Reese v. Western Union*.

57. Ibid.

58. Thomas Gaskell Shearman and Amasa Angell Redfield, *A Treatise on the Law of Negligence*, 4th ed. (New York: Baker, Voorhis, 1888), 379.

59. *So Relle v. Western Union*, 807–8.

60. Shearman and Redfield, *Treatise on the Law of Negligence*, 389.

61. Jabez Gridley Sutherland and John R. Berryman, *A Treatise on the Law of Damages, Embracing an Elementary Exposition of the Law, and Also Its Application to Particular Subjects of Contract and Tort*, 2nd ed., vol. 3 (Chicago: Callaghan, 1893), 2170.

62. Francis M. Burdick, "Tort Liability for Mental Disturbance and Nervous Shock," *Columbia Law Review* 5, no. 3 (1905): 191.

63. "Mental Anguish as Damage in Delayed Telegrams," *Yale Law Journal* 18, no. 5 (1909): 353.

64. Stearns, *American Cool*. For an important critique of Stearns's argument, see Barbara H. Rosenwein, "Worrying about Emotions in History," *American Historical Review* 107, no. 3 (2002): 821–45. Rosenwein offers two primary critiques of the "civilizing" assumptions she believes Stearns, following Norbert Elias, implies in his theories of emotionology. First, as a medievalist, Rosenwein objects to an implicit characterization of the medieval period as somehow implicitly *more emotional* than later periods. Second, and connected to the first point, she argues that claims about emotionology tend to flatten what is actually a varied and diverse community of feeling. Despite the usefulness of Rosenwein's critique, her reading of Stearns tends to oversimplify the emotionological approach, especially as it is articulated here. In explaining the approach to emotionology advocated in Stearns and Stearns's early work, Rosenwein argues that in the founding approach to emotionology the "emphasis, then, is not on how people felt or represented their feelings but on what people thought about such matters as crying in public, getting angry, or showing anger physically. It assumes that what people *think* about feelings they will eventually actually feel" (824). Here, Rosenwein strangely seems to equate "feeling" emotions with "representing" them, while separating "thinking about" feelings from both. What can it mean to "think about" emotions, if "representing" them is not a part of that thought? From the perspective of communication scholarship, it seems obvious that representing an emotion is part of the process of thinking about it, in that these representations become resources for making sense of emotional life. Likewise, I do not believe that the only consequence of such representations of emotion is whether or not people "eventually actually feel" them (and, in fact, I do

not see Stearns as assuming this). Regardless of whether or not specific individuals come to internalize a given representation or broader way of thinking about emotion, they may find themselves having to deal with it nonetheless. Even if a female lawyer working in a law firm does not believe that she is any more emotional than her male counterparts, she may find that she has to defend herself against claims of the hyperemotionality of women.

65. Peter Stearns, *American Fear: The Causes and Consequences of High Anxiety* (New York: Routledge, 2006), 104.

66. Linda W. Rosenzweig, *Another Self: Middle-Class American Women and Their Friends in the Twentieth Century* (New York: New York University Press, 1999).

67. Michael Barton, "Journalistic Gore: Disaster Reporting and Emotional Discourse in the *New York Times*, 1852–1956," in *An Emotional History of the United States*, ed. Peter Stearns and Jan Lewis (New York: New York University Press, 1998), 155–72. For other applications of Stearns's approach, see Jennifer Harding and E. Deidre Pribram, "Losing Our Cool?," *Cultural Studies* 18, no. 6 (2004): 863–83; Harding and Pribram, "The Power of Feeling: Locating Emotions in Culture," *European Journal of Cultural Studies* 5, no. 4 (2002): 407–26; Vanessa Pupavac, "War on the Couch: The Emotionology of the New International Security Paradigm," *European Journal of Social Theory* 7, no. 2 (2004): 149–70; and Stearns and Lewis, *Emotional History of the United States*.

68. Graham Wallas, *The Great Society: A Psychological Analysis* (New York: Macmillan, 1914), 3.

69. Walter Lippmann, *Public Opinion* (New York: Harcourt Brace, 1922), 56–57.

70. Walter Lippmann, *The Phantom Public* (New York: Harcourt Brace, 1925), 3.

71. John Dewey, *The Public and Its Problems* (New York: Holt, 1927), 98.

72. Gustave Le Bon, *The Crowd: A Study of the Popular Mind* (London: Unwin, 1897), 169.

73. Wallas, *Great Society*, 129.

74. Ibid., 132.

75. Lippmann, *Public Opinion*, 203.

76. "Motor Intoxication and Speed Madness," *New York Times*, July 21, 1903, 6.

77. "To Discourage Automobilists," *New York Times*, May 15, 1902, 8.

78. Kate Masterson, "The Monster in the Car," *Lippincott's Monthly Magazine*, August 1910, 204.

79. E. Favary, "The Evolution of the Automobile," *Phrenological Journal and Science of Health* 121, no. 1 (1908): 11.

80. "Stop! Look! Listen!," *New York Times*, June 16, 1905, 8.

81. "Better Car Drivers," *Los Angeles Times*, November 9, 1924, K12.

82. "Time to Slow Down for Cars and Demagogues," *Los Angeles Times*, October 17, 1906, II4.

83. William Augustus Evans, "How to Keep Well," *Washington Post*, May 1, 1923, 14.

84. "A Palace Car at Your Door—the Oldsmobile," *Town and Country*, February 11, 1905, 37; "Rambler," *New York Times*, January 15, 1905, 12, "The Stars of the Show," *Town and Country*, January 21, 1905, 5.

85. Veblen, *Theory of the Leisure Class* (see intro., n. 22).

86. "The Machine Gun—the World's Greatest Terror," *Popular Science Monthly*, October 1915, 458.

87. "War—on, beneath, and above the Waves," *Washington Post*, September 20, 1914, SM3. On the rhetorical negotiation of these images of death, see David Seitz, "Grave Negotiations: The Rhetorical Foundations of American World War I Cemeteries in Europe" (PhD diss., University of Pittsburgh, 2011).

88. "Shell Shock Psychic Result," *Los Angeles Times*, August 15, 1915, IV11; "New and Peculiar Military Cruelties Which Arise to Characterize Every War," *Washington Post*, May 30, 1915, SM4. For a discussion of the history of "shock" in the United States, see especially Schivelbusch, *The Railway Journey* (see intro., n. 31).

89. Edward M. Coffman, *The War to End All Wars: The American Military Experience in World War I* (Madison: University of Wisconsin Press, 1986).

90. Jill Frahm, "The Hello Girls: Women Telephone Operators with the American Expeditionary Forces during World War I," *Journal of the Gilded Age and Progressive Era* 3, no. 3 (2004).

91. "Station 'Hello Girls,'" *Great Lakes Recruit: A Pictorial Naval Magazine*, November 1918, 40.

92. George Washington Crile, *A Mechanistic View of War and Peace*, ed. Amy F. Rowland (New York: Macmillan, 1915).

93. David Marine, "George Washington Crile," *Science* 97, no. 2517 (1943): 277–78; Robert Hermann, "George Washington Crile (1864–1943)," *Journal of Medical Biography* 2 (1994): 78–83.

94. Otniel E. Dror, "Afterword: A Reflection on Feelings and the History of Science," *Isis* 100, no. 4 (2009): 850.

95. George Washington Crile, *An Experimental Research into Surgical Shock* (Philadelphia: Lippincott, 1899).

96. Crile, *A Mechanistic View of War and Peace*, 17.

97. Ibid., 11.

98. Ibid., 13.

99. Ibid., 49.

100. Ibid., 63.

101. Ibid., 58.

102. Ibid., 59, 99.

103. Ibid., 100.

104. Ibid., 103.

105. Lippmann, *Public Opinion*; Crile, *A Mechanistic View of War and Peace*, 103–4.

106. Ellis Wayne Hawley, *The Great War and the Search for a Modern Order: A History of the American People and Their Institutions, 1917–1933*, 2nd ed. (New York: St. Martin's, 1992), 7.

107. Ibid.; Ellis W. Hawley, "Herbert Hoover, the Commerce Secretariat, and the Vision of an 'Associative State,' 1921–1928," *Journal of American History* 61, no. 1 (1974): 116–40.

108. Thorstein Veblen, *The Engineers and the Price System* (New York: Huebsch, 1921).
109. William Henry Smyth, "'Technocracy'—National Industrial Management," *Industrial Management*, March 1919, 208. On the larger history of technocracy, including Veblen's place within it, see Donald R. Stabile, "Veblen and the Political Economy of the Engineer: The Radical Thinker and Engineering Leaders Came to Technocratic Ideas at the Same Time," *American Journal of Economics and Sociology* 45, no. 1 (1986): 41–52; William E. Akin, *Technocracy and the American Dream: The Technocrat Movement, 1900–1941* (Berkeley: University of California Press, 1977); Henry Elsner, *The Technocrats: Prophets of Automation* (Syracuse: Syracuse University Press, 1967); and Hawley, *Great War*.
110. American Engineering Council, *Waste in Industry* (Washington, DC: Federated American Engineering Societies, 1921).
111. Howard Scott, "Technocracy Speaks," *Living Age*, December 1932, 297. On Scott's importance to the technocracy movement, see Akin, *Technocracy and the American Dream*.
112. "Technocracy Idea Defended," *Los Angeles Times*, December 23, 1932, 2.
113. Hawley, *Great War*, 188–200.
114. American Engineering Council, *Waste in Industry*, 335, 65.
115. Smyth, "'Technocracy,'" 210.

NOTES TO CHAPTER 2

1. Elizabeth Eastlake, "Photography," *London Quarterly* 101, no. 202 (1857): 241–42.
2. Ibid., 253.
3. Allan Sekula, "The Body and the Archive," *October* 39 (1986): 4.
4. Walter Benjamin, "The Work of Art in the Age of Mechanical Reproduction," in *Illuminations* (New York: Harcourt Brace, 1968), 217–50. This essay originally appeared as "Das Kunstwerk im Zeitalter seiner technischen Reproduzierbarkeit" in 1936.
5. "Science for the People," *Eclectic Magazine of Foreign Literature*, May 1855, 48.
6. "A History of Photography in America," *Photographer's Friend*, January 1873, 7.
7. Charles Darwin, *The Expression of the Emotions in Man and Animals* (New York: Appleton, 1886), 355. For recent readings on Darwin's exploration of the emotions, see Peter J. Snyder et al., "Charles Darwin's Emotional Expression 'Experiment' and His Contribution to Modern Neuropharmacology," *Journal of the History of the Neurosciences* 19, no. 2 (2010): 158–70; Alison M. Pearn, "'This Excellent Observer . . . ': The Correspondence between Charles Darwin and James Crichton-Browne, 1869–75," *History of Psychiatry* 21, no. 2 (2010): 160–75; Daniel M. Gross, "Defending the Humanities with Charles Darwin's *The Expression of the Emotions in Man and Animals* (1872)," *Critical Inquiry* 37 (Autumn 2010): 34–59; Paul White, "Darwin's Emotions: The Scientific Self and the Sentiment of Objectivity," *Isis* 100, no. 4 (2009): 811–26; Ursula Hess and Pascal Thibault, "Darwin and Emotion Expression," *American Psychologist* 64, no. 2 (2009): 120–28; and Sarah Winter, "Darwin's Saussure: Biosemiotics and Race in Expression," *Representations* 107, no. 1 (2009): 128–61.

8. Francis Galton, "Composite Portraits, Made by Combining Those of Many Different Persons into a Single Resultant Figure," *Journal of the Anthropological Institute of Great Britain and Ireland* 8 (1879): 133. On Galton's use of photography, see Sekula, "The Body and the Archive," 3–64; and David Green, "Veins of Resemblance: Photography and Eugenics," *Oxford Art Journal* 7, no. 2 (1984): 3–16.

9. Galton, "Composite Portraits," 135.

10. For readings on the history of the stereoscope, see Judith Babbits, "Made in America: A Stereoscopic View of the United States," in *American Photographs in Europe*, ed. David Nye and Mick Gidley (Amsterdam: VU University Press, 1994), 41–56; Judith Babbitts, "Stereographs and the Construction of a Visual Culture in the United States," in *Memory Bytes: History, Technology, and Digital Culture*, ed. Lauren Rabinovitz and Abraham Geil (Durham: Duke University Press, 2004), 126–49; Helmut Gernsheim and Alison Gernsheim, *The History of Photography: From the Earliest Use of the Camera Obscura in the Eleventh Century up to 1914* (London: Oxford University Press, 1955); Jonathan Crary, *Techniques of the Observer* (see intro., n. 32); Heinz K. Henisch and Bridget Ann Henisch, *The Photographic Experience, 1839–1914: Images and Attitudes* (University Park: Pennsylvania State University Press, 1994); Robert Hirsch, *Seizing the Light: A History of Photography* (Boston: McGraw-Hill, 2000); Harold F. Jenkins, *Two Points of View: The History of the Parlor Stereoscope*, rev. ed. (Uniontown, PA: Warman, 1973); and Shirley Wajda, "A Room with a Viewer: The Parlor Stereoscope, Comic Stereographs, and the Psychic Role of Play in Victorian America," in *Hard at Play: Leisure in America, 1840–1940*, ed. Kathryn Grover (Amherst: University of Massachusetts Press, 1992), 112–38.

11. Ernst Mach, *Popular Scientific Lectures*, trans. Thomas J. McCormack (Chicago: Open Court, 1895), 73–74.

12. M. S. Emery, *Russia through the Stereoscope: A Journey across the Land of the Czar from Finland to the Black Sea* (New York: Underwood and Underwood, 1901), 27.

13. Gernsheim and Gernsheim, *History of Photography*, 191; Hirsch, *Seizing the Light*, 92.

14. Hirsch, *Seizing the Light*, 92.

15. Oliver Wendell Holmes, "The Stereoscope and the Stereograph," *Atlantic Monthly* 3, no. 20 (1859): 738–48.

16. Gernsheim and Gernsheim, *History of Photography*, 224.

17. David Brewster, *Letters on Natural Magic, Addressed to Sir Walter Scott* (London: Murray, 1834), 4.

18. David Brewster, "Spirit Rapping," *Christian Inquirer*, December 8, 1855, 4. Home related and described his exchange with Brewster in D. D. Home, *Incidents in My Life* (London: Pittman, 1864).

19. Brewster, *Letters on Natural Magic*, 7.

20. David Brewster, *The Stereoscope: Its History, Theory, and Construction, with Its Application to the Fine and Useful Arts and to Education* (London: John Murray, 1856), 180.

21. Ibid., 181.

22. "The Stereoscope," *Circular*, December 2, 1858, 180.

23. "Stereoscopic Pictures," *Southern Cultivator*, September 1860, 280.

24. "The Stereoscope," *Godey's Magazine and Lady's Book*, October 1852, 340.

25. Charles Baudelaire, *Art in Paris, 1845–1862*, trans. Jonathan Mayne (London: Phaidon, 1965), 153.

26. Harlan P. Beach, *Dawn on the Hills of T'ang* (New York: Student Volunteer Movement, 1898), 12; Harrison Burns, *Annotated Statutes of the State of Indiana, Showing the General Statutes in Force January 1, 1894* (Indianapolis: Bowen-Merrill, 1894), 682; "Ghostly Business," *Eclectic Magazine of Foreign Literature*, July 1863, 306.

27. Holmes, "The Stereoscope and the Stereograph," 744.

28. Gernsheim and Gernsheim, *History of Photography*, 193.

29. Hirsch, *Seizing the Light*, 94.

30. Frederick Winslow Taylor, *The Principles of Scientific Management* (New York: Harper and Brothers, 1911); Taylor, *Shop Management* (New York: Harper and Brothers, 1911).

31. Neil Postman, *Technopoly: The Surrender of Culture to Technology* (New York: Knopf, 1992), 50.

32. Martha Banta, *Taylored Lives: Narrative Productions in the Age of Taylor, Veblen, and Ford* (Chicago: University of Chicago Press, 1993), ix.

33. Ibid., 58.

34. Ibid.; Robert Kanigel, *The One Best Way: Frederick Winslow Taylor and the Enigma of Efficiency* (New York: Viking, 1997).

35. Albert E. Osborne, *The Stereograph and the Stereoscope, with Special Maps and Books Forming a Travel System: What They Mean for Individual Development, What They Promise for the Spread of Civilization* (New York: Underwood and Underwood, 1909), 25.

36. Daniel James Ellison, *Rome through the Stereoscope: Journeys in and about the Eternal City* (New York: Underwood and Underwood, 1902), xiii.

37. Joseph Mills Hanson, ed., *The World War through the Stereoscope: A Visualized, Vitalized History of the Greatest Conflict of All the Ages* (Meadville, PA: Keystone View, 1923), xx.

38. James Ricalton, *China through the Stereoscope: A Journey through the Dragon Empire at the Time of the Boxer Uprising* (New York: Underwood and Underwood, 1901), 11.

39. Crary, *Techniques of the Observer*, 123–24 (see intro., n. 32).

40. Hanson, *The World War through the Stereoscope*, xvii.

41. Mary Crawford, "The Laboratory Equipment of the Teacher of English," *English Journal* 4, no. 3 (1915): 149.

42. Mark Jefferson, "Stereoscopes in School," *Journal of Geography* 6, no. 5 (1907): 151, 153.

43. Francis Galton, "On Stereoscopic Maps, Taken from Models of Mountainous Countries," *Proceedings of the Royal Geographical Society of London* 9, no. 3

(1864): 104–5; Francis Galton and Galton Robert Cameron, "On Stereoscopic Maps, Taken from Models of Mountainous Countries," *Journal of the Royal Geographical Society of London* 35 (1865): 99–106.

44. F. Vivian Thompson, "Stereo-Photo Surveying," *Geographical Journal* 31, no. 5 (1908): 534–49; H. Hamshaw Thomas, "Geographical Reconnaissance by Aeroplane Photography, with Special Reference to the Work Done on the Palestine Front," *Geographical Journal* 55, no. 5 (1920): 349–70.

45. Jas. Mackenzie Davidson, "Remarks on the Value of Stereoscopic Photography and Skiagraphy: Records of Clinical and Pathological Appearances," *British Medical Journal* 2, no. 1979 (1898): 1669–71.

46. G. K. Gilbert, "Stereoscopic Study of the Moon," *Science* 13, no. 324 (1901): 407–9; David Waterston, *The Edinburgh Stereoscopic Atlas of Anatomy* (Edinburgh: Jack, 1906); Howard Riley Raper, *Elementary and Dental Radiography* (New York: Consolidated Dental, 1913); Alfred J. H. Iles, "Stereoscopic Radiography of Gunshot Wounds on Active Service," *British Medical Journal* 2, no. 2845 (1915): 54.

47. Frank Bunker Gilbreth, *Motion Sudy: A Method for Increasing the Efficiency of the Workman* (New York: Van Nostrand, 1911); Frank Bunker Gilbreth and Lillian Moller Gilbreth, *Applied Motion Study* (New York: Sturgis and Walton, 1917). On the Gilbreths' research, see Richard Lindstrom, "'They All Believe They Are Undiscovered Mary Pickfords,'" *Technology and Culture* 41, no. 4 (2000): 725–51; Sharon Corwin, "Picturing Efficiency: Precisionism, Scientific Management, and the Effacement of Labor," *Representations*, no. 84 (2003): 139–65; Nicholas Sammond, "Picture This: Lillian Gilbreth's Industrial Cinema for the Home," *Camera Obscura* 21, no. 63 (2006): 102–33; and Jon Hindmarsh, "Work and the Moving Image: Past, Present and Future," *Sociology* 43, no. 5 (2009): 990–96.

48. Edward Wheeler Scripture, "A Method of Stereoscopic Projection," *Scientific American* 73, no. 21 (1895): 327; E. W. Scripture, "Anaglyphs and Stereoscopic Projection," *Science* 10, no. 241 (1899): 185–87.

49. E. B. Titchener, "The Equipment of a Psychological Laboratory," *American Journal of Psychology* 11, no. 2 (1900): 258.

50. James Burt Miner, "A Case of Vision Acquired in Adult Life," *Psychological Review* (*University of Iowa Studies in Psychology*, no. 4) 6, no. 5 (1905): 105.

51. Ibid., 114.

52. "Nationally Known Psychologists Endorse the Stereograph," December 1933, folder 10-0/10-p, Keystone Travel Club, vol. 1, 1c, *Around the World with Burton Holes, America's Premier Traveler*, vol. 1. no. 7, JSSM.

53. "Book of Stereoscopic Technic," folder 10K, JSSM.

54. Crary, *Techniques of the Observer*, 124 (see intro., n. 32).

55. Osborne, *The Stereograph and the Stereoscope*, 26.

56. Jesse Lyman Hurlbut, *Traveling in the Holy Land through the Stereoscope: A Tour* (New York: Underwood and Underwood, 1905), 11 (emphasis added).

57. Ricalton, *China through the Stereoscope*, 16.

58. Osborne, *The Stereograph and the Stereoscope*, 27.

59. William James, "What Is an Emotion?," *Mind* 9, no. 34 (1884): 190.

60. Gerald Holton, "Ernst Mach and the Fortunes of Positivism in America," *Isis* 83, no. 1 (1992): 27–60.

61. James, "What Is an Emotion?," 203.

62. Ibid., 190.

63. William James, *The Principles of Psychology* (New York: Holt, 1890).

64. James, "What Is an Emotion?," 189.

65. Osborne, *The Stereograph and the Stereoscope*, 77.

66. Ibid., 78.

67. Ibid., 102.

68. Ibid., 125.

69. Ibid., 128.

70. U.S. Bureau of the Census, *Historical Statistics of the United States, Colonial Times to 1970, Bicentennial Edition* (Washington, DC: GPO, 1975), 14.

71. Ian F. Haney López, *White by Law: Legal Constructions of Race* (New York: New York University Press, 1996), 37–38 and passim.

72. Matthew Frye Jacobson, *Whiteness of a Different Color: European Immigrants and the Alchemy of Race* (Cambridge: Harvard University Press, 1998). For further discussions of whiteness in the United States, see Michael Rogin, *Blackface, White Noise: Jewish Immigrants in the Hollywood Melting Pot* (Berkeley: University of California Press, 1996); Thomas K. Nakayama and Judith N. Martin, *Whiteness: The Communication of Social Identity* (Thousand Oaks, CA: Sage, 1998); Mike Hill, ed., *Whiteness: A Critical Reader* (New York: New York University Press, 1997); and Richard Dyer, *White: Essays on Race and Culture* (London: Routledge, 1997).

73. On the rise of the managerial classes and middle management at the turn of the century, see Alfred Chandler, *The Visible Hand: The Managerial Revolution in American Business* (Cambridge: Belknap, 1977).

74. Melanie Archer and Judith Blau, "Class Formation in Nineteenth Century America: The Case of the Middle Class," *Annual Review of Sociology* 19 (1993): 26.

75. Taylor, *Principles of Scientific Management*, 102.

76. Osborne, *The Stereograph and the Stereoscope*, 127.

77. Ibid.

78. Ellen H. Richards, *Euthenics, the Science of Controllable Environment: A Plea for Better Living Conditions as a First Step toward Higher Human Efficiency* (Boston: Whitcomb and Barrows, 1910), vii. On the broader history of the euthenics movement, see Emma Seifrit Weigley, "It Might Have Been Euthenics: The Lake Placid Conferences and the Home Economics Movement," *American Quarterly* 26, no. 1 (1974): 79–96.

79. Richards, *Euthenics*, viii.

80. Carl E. Seashore, "Education for Democracy and the Junior College," *Bulletin of the American Association of University Professors* 13, no. 6 (1927): 399–404; Seashore, "The Term 'Euthenics,'" *Science* 94, no. 2450 (1941): 561–62; Seashore, "Origin of the Term 'Euthenics,'" *Science* 95, no. 2470 (1942): 455–56; Seashore, "The Scope

and Function of Euthenics," *Mental Hygiene* 33 (1949): 594–98; Brenton J. Malin, "Not Just Your Average Beauty: Carl Seashore and the History of Communication Research in the United States," *Communication Theory* 21, no. 3 (2011): 311–12.

81. Indeed, as Lauren Berlant has pointed out, the mythic status of Washington, DC, plays a central role in narratives of American citizenship. According to Berlant, "When Americans make the pilgrimage to Washington they are trying to grasp the nation in its totality." Lauren Gail Berlant, *The Queen of America Goes to Washington City: Essays on Sex and Citizenship* (Durham: Duke University Press, 1997), 25. "These secular pilgrimages," she further explains, "measure the intimate distance between the nation and some of the people who seek to be miraculated by its promise" (6).

82. Rufus Rockwell Wilson and Underwood and Underwood, *Washington through the Stereoscope: A Visit to Our National Capitol* (New York: Underwood and Underwood, 1904), 7–8.

83. Ibid., 36–37.

84. Ibid., 40–41.

85. Ibid., 47–48.

86. Keystone View Company, *Burial of the Victims of the* Maine *in Their Final Resting Place, Arlington Cemetery, Va., Dec. 28, 1899* (Meadville, PA: Keystone View, 1900), 1 photographic print on stereo card.

87. Wilson and Underwood and Underwood, *Washington through the Stereoscope*, 162.

88. Keystone View Company, *Doing Their Bit: Students of Mckinley Manual Training High School, Washington, D.C., with 4-Inch Shells They Have Made* (Meadville, PA: Keystone View, 1917), 1 photographic print on stereo card.

89. Keystone View Company, *Class in Wireless: National Service Camp for Girls, Washington, D.C. Outdoor Class* (Meadville, PA: Keystone View, 1917), 1 photographic print on stereo card.

90. Keystone View Company, *Setting Type on One Hundred Monotype Machines, Bureau of Printing and Engraving, Washington, D.C.* (Meadville, PA: Keystone View, 1917), 1 photographic print on stereo card.

91. Underwood and Underwood, *Pennsylvania Avenue from the Treasury, N.E. to the United States Capitol* (Washington, DC: Underwood and Underwood, 1903), graphic.

92. Fred Miller Robinson, *The Man in the Bowler Hat: His History and Iconography* (Chapel Hill: University of North Carolina Press, 1993); Anne Hollander, *Sex and Suits* (New York: Knopf, 1994); Chris Breward, "Manliness, Modernity and the Shaping of Male Clothing," in *Body Dressing*, ed. Joanne Entwistle and Elizabeth Wilson (Oxford: Berg, 2001), 165–81; Elizabeth Wilson, *Adorned in Dreams* (London: Virago, 1985).

93. For a discussion of the relationship between the stereoscope and the construction of the modern subject, see Suren Lalvani, *Photography, Vision, and the Production of Modern Bodies* (Albany: State University of New York Press, 1996), 176. According to Lalvani, the integration of the spectator with the "mechanical apparatus" of the stereoscope created a thoroughly modern conception of

viewing. "In general," he argues, "this model of vision constituted an observer who would be ideally suited to the increasing levels of mobility, exchange, and circulation of the image inherent to modernism's construction of the social."

94. Daniel James Ellison, *Italy through the Stereoscope: Journeys in and about Italian Cities* (New York: Underwood and Underwood, 1908), 66.

95. James Henry Breasted and Underwood and Underwood, *Egypt through the Stereoscope: A Journey through the Land of the Pharaohs* (New York: Underwood and Underwood, 1905), 126.

96. James Ricalton, *India through the Stereoscope: A Journey through Hindustan* (New York: Underwood and Underwood, 1907), 15.

97. Ellison, *Rome through the Stereoscope*, 314.

98. Ibid., 315 (emphasis added).

99. Ricalton, *China through the Stereoscope*, 107–8.

100. Keystone View Company, *A Block of Tenements in Which Some of China's Floating Population Dwell, Hong Kong, China* (Meadville, PA: Keystone View, 1906), 1 photographic print on stereo card.

101. "River Life in China," folder 10-0/10-p, Keystone Travel Club, vol. 1, *Keystone Travel Club*, vol. 2, no. 3, JSSM.

102. Keystone View Company, *Eskimo Girls in the Frigid Arctic, Cape York, Greenland* (Meadville, PA: Keystone View, n.d.), 1 photographic print on stereo card.

103. Keystone View Company, *A Native New Guinea Family and Its Pets* (Meadville, PA: Keystone View, 1919), 1 photographic print on stereo card.

104. Keystone View Company, *White Robed Pottery Peddlers on the Streets of Seoul, Korea* (Meadville, PA: Keystone View, 1901), 1 photographic print on stereo card.

105. Keystone View Company to Walter Barnhart, June 30, 1902, Letters to Salesman Walter W. Barnhart, folder 3A-3J, JSSM.

106. Underwood and Underwood, *Canvass and Delivery: The Underwood Travel System* (New York: Underwood and Underwood, 1909), 6; Keystone View Company to Walter Barnhart, August 20, 1902, folder 3A-3J, Letters to Salesman Walter W. Barnhart, JSSM.

107. Keystone View Company to Walter Barnhart, September 26, 1902, folder 3A-3J, Letters to Salesman Walter W. Barnhart, JSSM.

108. Underwood and Underwood, *Canvass and Delivery*, 52–53.

109. "Greetings," September 1931, folder 10-0/10-p, Keystone Travel Club, vol. 1, 2c, *Keystone Travel Club*, vol. 1, no. 1, JSSM.

110. "Where Have You Traveled? What Have You Seen?," September 1931, folder 10-0/10-p, Keystone Travel Club, vol. 1, 2c, *Keystone Travel Club*, vol. 1, no. 1, JSSM.

111. "Something Really Different," November 1931, folder 10-0/10-p, Keystone Travel Club, vol. 1, 1c, *Around the World with Burton Holes, America's Premier Traveler*, vol. 1, no. 6, JSSM.

112. "Final Authority at Hollywood," November 1931, folder 10-0/10-p, Keystone Travel Club, vol. 1, 1c, *Around the World with Burton Holes, America's Premier Traveler*, vol. 1, no. 6, JSSM.

113. "Letter from Betty Lou Hall to Anne Travelog," March 1932, folder 10-0/10-p, Keystone Travel Club, vol. 1, 2c, *Keystone Travel Club*, vol. 2, no. 3, JSSM.

114. Stephen Arnold Goldstein, "A Trip round the World," August 1931, folder 10-0/10-p, Keystone Travel Club, vol. 1, 1c, *Around the World with Burton Holes, America's Premier Traveler*, vol. 1, no. 3, JSSM.

115. Lillian Bell, "My Tour," September 1931, folder 10-0/10-p, Keystone Travel Club, vol. 1, 2c, *Keystone Travel Club*, vol. 1, no. 1, JSSM.

116. June Margaret Tracy, "Eskimos," December 1931, folder 10-0/10-p, Keystone Travel Club, vol. 1, 2c, *Keystone Travel Club*, vol. 1, no. 3, JSSM.

117. Keystone View Company, *Visual Education through Stereographs and Lantern Slides* (Meadville, PA: Keystone View, 1917).

NOTES TO CHAPTER 3

1. Virgil Pinkley, *Essentials of Elocution and Oratory* (New York: Werner, 1897), 17.

2. Thomas Edison, "The Perfected Phonograph," *North American Review* 379, no. 146 (1888): 641.

3. Herman Cohen, *The History of Speech Communication: The Emergence of a Discipline, 1914–1945* (Annandale, VA: Speech Communication Association, 1994), 13.

4. Ibid., 12.

5. Kenneth Cmiel, *Democratic Eloquence: The Fight over Popular Speech in Nineteenth-Century America* (Berkeley: University of California Press, 1991), 13–14.

6. Leff and Procario, "Rhetorical Theory in Speech Communication," 4 (see chap. 1, n. 38).

7. Walker, *Elements of Elocution*, 264 (see chap. 1, n. 43).

8. Ibid., 305.

9. Shoemaker, *Advanced Elocution*, 15 (see chap. 1, n. 40).

10. Ibid., 110.

11. Smiley Blanton, "The Voice and the Emotions," *Quarterly Journal of Public Speaking* 1, no. 2 (1915): 162–63.

12. Kathleen O'Keeffe, *Charm of the Spoken Word* (Huntington, IN: OSV, 1929), 47–48.

13. F. H. Lane, "Action and Emotion in Speaking," *Quarterly Journal of Public Speaking* 2, no. 3 (1916): 227.

14. "Annual Catalogue of the Emerson College and the Boston School of Oratory," 1921, ECSC; "Annual Catalogue of the Emerson College and the Boston School of Oratory," 1922, ECSC; "Annual Catalogue of the Emerson College and the Boston School of Oratory," 1919, ECSC.

15. Wayland Parrish, "The Style of Extemporaneous Speech," *Quarterly Journal of Speech Education* 9, no. 4 (1923): 346.

16. Dale Carnegie, *Public Speaking: A Practical Course for Business Men* (New York: Association Press, 1926), 144.

17. Edward Z. Rowell, "Public Speaking in a New Era," *Quarterly Journal of Speech* 16, no. 1 (1930): 62.

18. Lawrence B. Goodrich, "The Illusion of Real Talk," *Quarterly Journal of Speech* 19, no. 1 (1933): 39.

19. Leff and Procario, "Rhetorical Theory in Speech Communication," 5–12 (see chap. 1, n. 38).

20. James A. Winans, "Public Speaking I at Cornell University," *Quarterly Journal of Public Speaking* 3 (1917): 154.

21. Hoyt H. Hudson, "*A Course of Lectures on Oratory and Criticism* by Joseph Priestly," *Quarterly Journal of Speech Education* 13, no. 1 (1927): 89.

22. Ernest Keen, *A History of Ideas in American Psychology* (Westport, CT: Praeger, 2001); Deborah Coon, "Standardizing the Subject: Experimental Psychologists, Introspection, and the Quest for a Technoscientific Ideal," *Technology and Culture* 34, no. 4 (1993): 757–83; Charles Bazerman, *Shaping Written Knowledge: The Genre and Activity of the Experimental Article in Science* (Madison: University of Wisconsin Press, 1988); Stephen Toulmin and David Leary, "The Cult of Empiricism in Psychology and Beyond," in *A Century of Psychology as Science,* ed. Sigmund Koch and David E. Leary (New York: McGraw-Hill, 1985).

23. Mark C. Smith, *Social Science in the Crucible: The American Debate over Objectivity and Purpose, 1918–1941* (Durham: Duke University Press, 1994); Charles Camic, "On Edge: Sociology during the Great Depression and the New Deal," in *Sociology in America: A History,* ed. Craig J. Calhoun (Chicago: University of Chicago Press, 2007), 225–80; Robert C. Bannister, *Sociology and Scientism: The American Quest for Objectivity, 1880–1940* (Chapel Hill: University of North Carolina Press, 1987).

24. Otniel Dror, "Techniques of the Brain and the Paradox of Emotions, 1880–1930," *Science in Context* 14, no. 4 (2001): 643–61; Dror, "Counting the Affects: Discoursing in Numbers," *Social Research* 68, no. 2 (2001): 357–79; Dror, "The Affect of Experiment: The Turn to Emotions in Anglo-American Physiology, 1900–1940," *Isis* 90, no. 2 (1999): 205–38; Dror, "The Scientific Image of Emotion: Experience and Technologies of Inscription," *Configurations* 7, no. 3 (1999): 355–401.

25. E. W. Scripture, "Speech Curves," *Modern Language Notes* 16, no. 3 (1901): 73.

26. Edward Wheeler Scripture, "How the Voice Looks," *Century Illustrated Magazine* 64, no. 1 (1902): 168.

27. Matataro Matsumoto, "Researches on Acoustic Space," *Studies from the Yale Psychological Lab* 5 (1897): 1–75.

28. Titchener, "Equipment of a Psychological Laboratory" (see chap. 2, n. 49); Edward Bradford Titchener, *Experimental Psychology: A Manual of Laboratory Practice* (New York: Macmillan, 1901).

29. For a more detailed discussion of the work of Carl Seashore and his place in the larger history of communication research, see Malin, "Not Just Your Average Beauty" (see chap. 2, n. 80).

30. Seashore, "A Voice Tonoscope," 19 (see intro., n. 28).

31. Seashore, "Phonophotography," 466 (see intro., n. 28).

32. Carl E. Seashore and Milton Metfessel, "Deviation from the Regular as an Art Principle," *Proceedings of the National Academy of Sciences of the United States of America* 11, no. 9 (1925): 538.

33. Carl E. Seashore, "The Natural History of the Vibrato," *Proceedings of the National Academy of Sciences* 17, no. 12 (1931): 623.

34. Glenn N. Merry, "Notes from the Classroom and Laboratory," *Quarterly Journal of Speech Education* 7, no. 2 (1921): 172.

35. Glenn N. Merry, "Voice Inflection in Speech," *Psychological Monographs* 31, no. 1 (1922): 205.

36. Glenn Newton Merry, "A Roentgenological Method of Measuring the Potentiality of Voice Resonance," *Quarterly Journal of Speech Education* 5, no. 1 (1919): 27.

37. Glenn Newton Merry, "Accessory Sinuses and Head Resonance," *Quarterly Journal of Public Speaking* 3, no. 3 (1917): 274.

38. G. Oscar Russell, *The Vowel: Its Physiological Mechanisms as Shown by X-Ray* (Columbus: Ohio State University Press, 1928), 257.

39. C. M. Wise, "'Chest Resonance' (a Report of Work in Progress)," *Quarterly Journal of Speech* 18, no. 3 (1932): 451.

40. "Department of English, Bulletin of the University of Wisconsin," 1905–1906, series 7/10/00-1, box 1, Department of English, DAUW; "Department of English, Bulletin of the University of Wisconsin," 1907–1908, series 7/10/00-1, box 1, Department of English, DAUW.

41. James O'Neill to E. A. Birge, February 26, 1914, series 7/35/15, box 1, folder: O'Neill, A–H, 1914–1919, Department of Speech, DAUW.

42. Smiley Blanton, "Research Problems in Voice and Speech," *Quarterly Journal of Public Speaking* 2 (1916): 9–17.

43. Andrew Thomas Weaver to Thomas Edison Inc., December 15, 1919, series 7/35/11, box 1, folder: Weaver, A–G, 1919–1925, Papers of Andrew T. Weaver, DAUW.

44. Andrew Thomas Weaver to Western Electric Company, March 17, 1925, series 7/35/11, box 1, folder: Weaver, M–Z, 1919–1925, Papers of Andrew T. Weaver, DAUW; Andrew Thomas Weaver to Denoyer-Geppert Company, October 1, 1925, series 7/35/2, box 1, Apparatus, 1924–1925, Department of Speech, DAUW.

45. Andrew Thomas Weaver, "Experimental Studies in Vocal Expression," *Quarterly Journal of Speech Education* 10, no. 3 (1924): 199.

46. "Admissions Papers of Robert West," 1919–1925, DAUW; "Faculty Card of Robert West," DAUW.

47. Robert West, "A Commercial Device Appraised," *Quarterly Journal of Speech Education* 10, no. 4 (1924): 390.

48. Robert West, "Notes on Apparatus Usable in the Study of Voice," *Quarterly Journal of Speech Education* 11, no. 3 (1925): 243.

49. Robert West, "The Nature of Vocal Sounds," *Quarterly Journal of Speech Education* 7, no. 4 (1926): 252.

50. Sterne, *The Audible Past*, 40 (see intro., n. 30).

51. Robert J. Copeland and Christian H. Stoelting, "Autographic Recording Device," U.S. Patent 513,558, filed January 26, 1893, issued January 30, 1894.

52. Christian H. Stoelting, "Roll Holding Camera," U.S. Patent 533,618, filed March 24, 1894, issued February 5, 1895.

53. "Christian H. Stoelting," *American Journal of Psychology* 56, no. 3 (1943): 450; Christian H. Stoelting, "Barometer," U.S. Patent 756,905, filed August 15, 1903, issued April 12, 1904; Christian H. Stoelting, "Laboratory-Weights," U.S. Patent 756,905, filed August 15, 1903, issued April 25, 1905.

54. Walter Miles, *Carl Emil Seashore, 1866–1949: A Biographical Memoir* (Washington, DC: National Academy of Sciences, 1956). On Seashore's place in the development of the audiometer, as well as the history of the audiometer more generally, see Jonathan Sterne, *Mp3: The Meaning of a Format* (Durham: Duke University Press, 2012).

55. Carl E. Seashore, *Manual of Instructions and Interpretations for Measures of Musical Talent*, State University of Iowa Standard Tests (Chicago: Stoelting, n.d.), 1, Papers of Carl Seashore (RG 99.0164), box 2, folder M, UAUI.

56. George D. Stoddard, "Carl Emil Seashore: 1866–1949," *American Journal of Psychology* 63, no. 3 (1950): 456–62; Miles, *Carl Emil Seashore*; "New Tests Aid in Music," *New York Times*, December 2, 1934, XX8.

57. Seashore, *Manual of Instructions*, 2.

58. Smiley Blanton, Margaret Blanton, and Sara Stinchfield, *Blanton-Stinchfield Speech Measurements: Devised in the Speech Clinic of the Department of Speech at the University of Wisconsin* (Chicago: Stoelting, 1926).

59. "Faculty Card of E. Ray Skinner," DAUW; "Admissions Papers of E. Ray Skinner," 1919–1925, DAUW.

60. Andrew Thomas Weaver to John Mills, November 22, 1926, series 7/35/11-1, box 1, folder: Weaver, 1925–27, Papers of Andrew T. Weaver, DAUW.

61. E. Ray Skinner, "A Calibrated Recording and Analysis of the Pitch, Force and Quality of Vocal Tones Expressing Happiness and Sadness," *Speech Monographs* 2, no. 1 (1935): 81–137.

62. Gladys Borchers, "Interview #9," University of Wisconsin, Madison Oral History Project, DAUW.

63. "Faculty Card of Edward Hall Gardner," DAUW; "Class Breaks Record," *Daily Cardinal*, February 26, 1921; J. W. Cunliffe to Dean Birge, January 6, 1910, series 7/1/2-1, box 4, Birge Correspondence Budget Papers, July 1909–March 1911, College of Letters and Science, Administration Dean's Files, University Archives, Steenbrock Library, University of Wisconsin, Madison; J. W. Cunliffe to Dean Birge, February 24, 1911, series 7/1/2-1, box 4, Birge Correspondence Budget Papers, July 1909–March 1911, College of Letters and Science, Administration Dean's Files, University Archives, Steenbrock Library, University of Wisconsin, Madison; W. A. Scott to Dean Sellery, January 27, 1919, series 7/1/4-2, box 1, Dean Sellery, University Archives, Steenbrock Library, University of Wisconsin, Madison; W. A. Scott to J. W.

Cunliffe, October 20, 1910, series 7/1/2-1, box 4, Birge Correspondence Budget Papers, July 1909–March 1911, College of Letters and Science, Administration Dean's Files, University Archives, Steenbrock Library, University of Wisconsin, Madison.

64. Edward Hall Gardner, *Effective Business Letters: Their Requirements and Preparation, with Specific Directions for the Various Types of Letters Commonly Used in Business* (New York: Ronald Press, 1915), 7–8.

65. Ibid., iii.

66. "Are You Embarassed by Mistakes in Pronunciation?," *Los Angeles Times*, March 4, 1928, K25.

67. "Cultured Speech—by a New Method," *English Journal* 20, no. 10 (1931): 876.

68. "'Faux Pas' I Said . . . And Everyone Tittered," *New McClure's* 61, no. 5 (1928): 13.

69. "Are You Embarrassed by Mistakes in Pronunciation?," *Forum*, May 1928, 8.

70. "'Faux Pas' I Said."

71. Edward Hall Gardner and E. Ray Skinner, *Good Taste in Speech: The Manual of Instruction of the Pronunciphone Course* (Chicago: Pronunciphone, 1928), 7.

72. E. Ray Skinner, *Pronunciphone* (Chicago: Pronunciphone, 1920), sound recording.

73. Norman Charles Meier, "A Measure of Art Talent," *Psychological Monographs* 39, no. 2 (1928): 184–99; Norman Charles Meier and Carl E. Seashore, *The Meier-Seashore Art Judgment Test: Examiner's Manual* (Iowa City: Bureau of Educational Research and Service, 1930).

74. "Body Is Medium for Radio Waves Always in Air," *Atlanta Constitution*, July 2, 1922, A6. On the eeriness of radio waves in the 1920s, see John Durham Peters, "The Uncanniness of Mass Communication in Interwar Social Thought," *Journal of Communication* 46, no. 3 (1996): 108–24.

75. Carl Dreher, "Communized Emotion," *Forum* 74, no. 6 (1925): 844.

76. Paddy Scannell and David Cardiff, *A Social History of British Broadcasting*, vol. 1, *1922–1939: Serving the Nation* (Oxford, UK: Basil Blackwell, 1991), 162.

77. Douglas, *Listening In*, 102 (see intro., n. 30).

78. "Qualifications Necessary to Be a Radio Announcer," *New York Times*, November 16, 1924, XX17.

79. "Effort Made to Outline Announcer's Technique," *Christian Science Monitor*, February 12, 1925, 16; "Study of Voice Technique to Aid Radio Announcers," *New York Times*, February 15, 1925, XX14.

80. "Committee Outlines 'Ideal' Radio Announcer's Voice," *Christian Science Monitor*, February 16, 1925, 13.

81. "Announcers in East Tested by New Standard," *Chicago Daily Tribune*, March 8, 1925, D12.

82. Barnouw, *A Tower in Babel* (see intro., n. 30).

83. Richard Aldrich, "Graham McNamee's Song Recital," *New York Times*, October 23, 1920, 17; Agnes Smith, "This Is Graham McNamee," *Life*, June 14, 1928, 15.

84. Smith, "This Is Graham McNamee," 15.

85. Paula Wera, "The Listener's Viewpoint," *Washington Post*, August 19, 1928, R7.

86. "Personality on the Air," *New York Times*, March 30, 1932, X14.

87. Grenville Kleiser, *Radio Broadcasting: How to Speak Convincingly* (New York: Funk and Wagnalls, 1935), 41.

88. Donald Horton and Richard Wohl, "Mass Communication and Para-Social Interaction: Observations on Intimacy at a Distance," *Psychiatry* 19 (1956): 215–29.

89. "Pick Men as Best Radio Announcers," *New York Times*, July 29, 1926, 16.

90. Sherman P. Lawton, "The Principles of Effective Radio Speaking," *Quarterly Journal of Speech* 16, no. 3 (1930): 270.

91. Ralph L. Power, "Picking Best Announcer Hard Task for Radio," *Los Angeles Times*, March 24, 1929, D8.

92. Henry Adams Bellows, "Broadcasting and Speech Habits," *Quarterly Journal of Speech* 17, no. 2 (1931): 246.

93. "'Honest Voice' of Radio Announcer Wins Confidence of Stock Investor," *Christian Science Monitor*, January 9, 1933, 3.

94. "Examination for Announcers," *New York Times*, May 4, 1930, XX19.

95. "Some Tongue-Twisting Tests," *Christian Science Monitor*, April 26, 1932, 4.

96. "Only 10 of 2,500 Applicants Pass Radio Announcers' Test," *New York Times*, December 15, 1931, 12.

97. "Gesture Is Necessary in Radio Broadcasting Says Mrs. Puffer," *Emerson College News* 5, no. 7 (1926).

98. Mildred Holland, "Making the Most of Your Personality: Study Speech by Means of Radio," *Washington Post*, April 10, 1924, 20.

99. Charles Henry Woolbert and Andrew Thomas Weaver, *Better Speech: A Textbook of Speech Training for Secondary Schools* (New York: Harcourt, Brace, 1929), 71.

100. Allison McCracken, "'God's Gift to Us Girls': Crooning, Gender, and the Re-Creation of American Popular Song, 1928–1933," *American Music* 17, no. 4 (1999): 365–95.

101. Someone referring to himself or herself simply as Emolas expressed this concern in a letter to the *Washington Post* in 1934. Emolas, "He Agrees with Mr. Ayers," *Washington Post*, January 14, 1934, 8. Emolas wrote in agreement with a letter by Grover Ayers identifying the assault on "the King's English" by bad radio voices. In addition to discussing other problems, Ayers commented on the problem of crooning and other popular singing. "Just why the housewife is thought to want to hear some girl and boy sing through his or her nose about his or her tale of woe is beyond me," he wrote. Grover Ayers, "The King's English," *Washington Post*, January 11, 1934, 6.

102. John Carlisle, "Priest of a Parish of the Air Waves," *New York Times*, October 29, 1933, SM8. On Father Coughlin's complicated time on the airwaves, see Alan Brinkley, *Voices of Protest: Huey Long, Father Coughlin, and the Great Depression* (New York: Vintage, 1983).

103. "Personality on the Air."

104. "Brevity and Appeal to Reason Make Radio Talks Magnetic," *New York Times*, March 25, 1928.

105. "Seek to Make Distinct Science of Announcing," *Chicago Daily Tribune*, February 15, 1925, D8.

106. Richard C. Borden and Alvin Clayton Busse, *Speech Correction* (New York: Crofts, 1925); Borden and Busse, *How to Win an Argument* (New York: Harper and Brothers, 1926); Borden and Busse, *The New Public Speaking* (New York: Harper and Brothers, 1930).

107. Henry Lee Ewbank, "Studies in the Techniques of Radio Speech," *Quarterly Journal of Speech* 18, no. 4 (1932): 560.

108. Frederick Hillis Lumley, *Broadcasting Foreign-Language Lessons*, Ohio State University Studies (Columbus: Ohio State University, 1934); Lumley, "Rates of Speech in Radio Speaking," *Quarterly Journal of Speech* 19, no. 3 (1933): 393–403; Lumley, "Does Radio Broadcasting Help Pupils Pronounce a Foreign Language?," *Modern Language Journal* 18, no. 6 (1934): 383–88.

109. "Institute for Radio Education," 1930–1938, container 52, folder 988, PFIR.

110. "Annual Catalogue of the Emerson College and the Boston School of Oratory," 1932, ECSC; "Radio Address Course Opens with Edes as Instructor," *Emerson College News* 11, no. 5 (1932).

111. "Study of Broadcasting Is Offered at Michigan," *New York Times*, July 8, 1934, XX6.

112. Ray K. Immel, "Speech and the Talking Pictures," *Quarterly Journal of Speech* 15, no. 2 (1929): 159–65; "Colleges Offer Courses in Broadcast Speech and Microphone Technique," *New York Times*, April 2, 1933, X8.

113. Peters, "Uncanniness of Mass Communication," 109.

114. H. Twitchell, "How Shall We Talk?," *Education* 28, no. 2 (1907): 115.

115. Ibid., 118.

116. Terese Rose Nagel, "Shall We Have Women Announcers?," *Christian Science Monitor*, March 10, 1931, 5.

117. "Radio Reveals Voices," *Washington Post*, May 15, 1927, SM8.

118. Nagel, "Shall We Have Women Announcers?"

119. "Pick Men as Best Radio Announcers."

120. Nagel, "Shall We Have Women Announcers?"

121. "Female Announcers for New Tokyo Radio," *Washington Post*, August 23, 1925, AU2.

122. Katherine Ward Fisher, "Women Radio Announcers," *New York Times*, May 24, 1926, 18.

123. Herbert Glover, "Honor Won by Woman over Radio," *Los Angeles Times*, June 12, 1927, 12.

124. R. P. Jutson, "Women's Voices on the Radio," *New York Times*, May 28, 1926, 20.

125. Ruth Oldenziel, *Making Technology Masculine: Men, Women and Modern Machines in America, 1870–1945* (Amsterdam: Amsterdam University Press, 1999). On the question of "the useful arts," see also Nina E. Lerman, "The Uses of Useful Knowledge: Science, Technology, and Social Boundaries in an Industrializing City," *Osiris* 12 (1997): 39–59; and Ronald Kline, "Construing 'Technology' as

'Applied Science': Public Rhetoric of Scientists and Engineers in the United States, 1880–1945," *Isis* 86, no. 2 (1995): 194–221. A number of scholars have taken up the question of the gendered nature of technology as well, including Nina E. Lerman, Arwen Palmer Mohun, and Ruth Oldenziel, "Versatile Tools: Gender Analysis and the History of Technology," *Technology and Culture* 38, no. 1 (1997): 1–8; Lerman, Mohun, and Oldenziel, "The Shoulders We Stand on and the View from Here: Historiography and Directions for Research," *Technology and Culture* 38, no. 1 (1997): 9–30; Judy Wajcman, *Feminism Confronts Technology* (Cambridge: Polity, 1991); Wajcman, "Reflections on Gender and Technology Studies: In What State Is the Art?," *Social Studies of Science* 30, no. 3 (2000): 447–64; Judith A. McGaw, "Inventors and Other Great Women: Toward a Feminist History of Technological Luminaries," *Technology and Culture* 38, no. 1 (1997): 214–31; McGaw, "Women and the History of American Technology," *Signs* 7, no. 4 (1982): 798–828; Ruth Schwartz Cowan, "The 'Industrial Revolution' in the Home: Household Technology and Social Change in the 20th Century," *Technology and Culture* 17, no. 1 (1976): 1–23; Cowan, "From Virginia Dare to Virginia Slims: Women and Technology in American Life," *Technology and Culture* 20, no. 1 (1979): 51–63; Joan Rothschild, "A Feminist Perspective on Technology and the Future," *Women's Studies International Quarterly* 4, no. 1 (1981): 65–74; Knut H. Sørensen, "Towards a Feminized Technology? Gendered Values in the Construction of Technology," *Social Studies of Science* 22, no. 1 (1992): 5–31.

126. Rotundo, *American Manhood* (see chap. 1, n. 37); Bederman, *Manliness and Civilization* (see chap. 1, n. 37); Kimmel, *Manhood in America* (see chap. 1, n. 37).

127. Douglas, *Inventing American Broadcasting*, 191 (see intro., n. 30).

128. C. L. Allen, "A Simple Ready-Made Aerial," *Wireless Age* 7, no. 6 (1920): 33.

129. Douglas, *Inventing American Broadcasting*, 188–94 (see intro., n. 30).

130. Jaggar, "Love and Knowledge," 152 (see intro, n. 18).

131. "The Haunting Charm of Hawaiian Music," *New York Times*, August 1, 1916, 5 (all emphasis in original). Significantly, Columbia Records was also involved in helping Columbia University professors Harry Morgan Ayres and W. Cabell Greet create an archive of "American speech records." The two professors traveled the country recording examples of American speech, and then released these on Columbia Records for future analysis. Harry Morgan Ayres and W. Cabell Greet, "American Speech Records at Columbia University," *American Speech* 5, no. 5 (1930): 333–58.

132. Milton Metfessel, "The Collecting of Folk Songs by Phonophotography," *Science* 67, no. 1724 (1928): 28.

133. Ibid., 30.

134. J. Walter Fewkes, "On the Use of the Phonograph in the Study of the Languages of American Indians," *Science* 15, no. 378 (1890): 267.

135. Frances Densmore, "Scale Formation in Primitive Music," *American Anthropologist* 11, no. 1 (1909): 1–12.

136. Charles K. Wead, "The Study of Primitive Music," *American Anthropologist* 2, no. 1 (1900): 75–79; E. Sapir, "Percy Grainger and Primitive Music," *American*

Anthropologist 18, no. 4 (1916): 592–97; Helen H. Roberts, "Melodic Composition and Scale Foundations in Primitive Music," *American Anthropologist* 34, no. 1 (1932): 79–107; Erwin Felber and Theodore Baker, "New Approaches to Primitive Music: The Music of Infants; Neurasthenics and Paranoiacs; Drunkards and Deafmutes," *Musical Quarterly* 19, no. 3 (1933): 288–302; George Herzog, "Speech-Melody and Primitive Music," *Musical Quarterly* 20, no. 4 (1934): 451–66; Frances Densmore, "The Study of Indian Music in the Nineteenth Century," *American Anthropologist* 29, no. 1 (1927): 77–86; Densmore, "What Intervals Do Indians Sing?," *American Anthropologist* 31, no. 2 (1929): 271–76; Densmore, "Peculiarities in the Singing of the American Indians," *American Anthropologist* 32, no. 4 (1930): 651–60.

137. Ray Miller, "A Strobophotographic Analysis of a Tlingit Indian's Speech," *International Journal of American Linguistics* 6, no. 1 (1930): 47.

138. C. M. Wise, "Negro Dialect," *Quarterly Journal of Speech* 19, no. 4 (1933): 527.

139. Judith Butler, *Bodies That Matter: On the Discursive Limits of "Sex"* (New York: Routledge, 1993), 48–49.

140. Carl E. Seashore, "Three New Approaches to the Study of Negro Music," *Annals of the American Academy of Political and Social Science* 140 (1928): 191.

NOTES TO CHAPTER 4

1. William Marston to John A. Larson, April 5, 1936, box 1, folder 13, JALP.

2. On the connections between Marston's lie detection and his creation of Wonder Woman, see Geoffrey Bunn, "The Lie Detector, Wonder Woman and Liberty: The Life and Work of William Moulton Marston," *History of the Human Sciences* 10, no. 1 (1997): 91–119.

3. Henry James Forman, *Our Movie Made Children* (New York: Macmillan, 1933), 4.

4. Ibid., 1.

5. Wendell Dysinger and Christian Ruckmick, *The Emotional Responses of Children to the Motion Picture Situation* (New York: Macmillan, 1933).

6. Ibid., 6.

7. Lisa Cartwright, "'Experiments of Destruction': Cinematic Inscriptions of Physiology," *Representations* 40 (1992): 129–52; Merrily Borell, "Extending the Senses: The Graphic Method," *Medical Heritage* 2 (1986): 114–21; Marta Braun, "The Photographic Work of E. J. Marey," *Studies in Visual Communication* 9 (1983): 4–23.

8. Otniel Dror, "Creating the Emotional Body: Confusion, Possibilities and Knowledge," in Stearns and Lewis, *An Emotional History of the United States*, 173 (see chap. 1, n. 67).

9. Timothy Lenoir, "Helmholtz and the Materialities of Communication," *Osiris* 9 (1994): 185–207; John Durham Peters, "Helmholtz, Edison, and Sound History," in Rabinovitz and Geil, *Memory Bytes*, 177–98 (see chap. 2, n. 10); Sterne, *The Audible Past* (see intro., n. 30).

10. Kurt Danziger, *Constructing the Subject: Historical Origins of Psychological Research* (Cambridge: Cambridge University Press, 1990), 37.

11. Ibid., 39.

12. Christian A. Ruckmick, "Carl Stumpf," *Psychological Bulletin* 34, no. 4 (1937): 187.

13. Alexandra Hui, "Hearing Sound as Music: Psychophysical Studies of Sound Sensation and the Music Culture of Germany, 1860–1910" (PhD diss., University of California, 2008). On Wundt more generally, see Malcolm Ashmore, Steven D. Brown, and Katie Macmillan, "Lost in the Mall with Mesmer and Wundt: Demarcations and Demonstrations in the Psychologies," *Science, Technology and Human Values* 30, no. 1 (2005): 76–110; Ludy T. Benjamin, *A Brief History of Modern Psychology* (Malden, MA: Blackwell, 2007); Ruth Benschop and Douwe Draaisma, "In Pursuit of Precision: The Calibration of Minds and Machines in Late Nineteenth-Century Psychology," *Annals of Science* 57, no. 1 (2000): 1–25; Wan-chi Wong, "Retracing the Footsteps of Wilhelm Wundt: Explorations in the Disciplinary Frontiers of Psychology and in Völkerpsychologie," *History of Psychology* 12, no. 4 (2009): 229–65; R. W. Rieber and David Kent Robinson, *Wilhelm Wundt in History: The Making of a Scientific Psychology* (New York: Kluwer Academic/Plenum, 2001).

14. Lorraine Daston and Peter Galison, "The Image of Objectivity," *Representations* 40 (Special Issue: Seeing Science) (1992): 81.

15. Ibid.

16. Ibid., 83.

17. Smith, *Social Science in the Crucible*, 28 (see chap. 3, n. 23).

18. Dror, "Scientific Image of Emotion," 364–67 (see chap. 3, n. 24).

19. Ibid., 358.

20. George Trumbull Ladd, *Psychology, Descriptive and Explanatory: A Treatise of the Phenomena, Laws, and Development of Human Mental Life* (New York: Scribner's, 1894), 17.

21. Danziger, *Constructing the Subject*, 42–48; Keen, *History of Ideas in American Psychology* (see chap. 3, n. 22); Dror, "Scientific Image of Emotion" (see chap. 3, n. 24); Bazerman, *Shaping Written Knowledge* (see chap. 3, n. 22); Toulmin and Leary, "Cult of Empiricism in Psychology" (see chap. 3, n. 22); Coon, "Standardizing the Subject," 757–83 (see chap. 3, n. 22). According to both Bazerman and Keen, John Watson's behaviorism also played a role in this movement away from philosophical inquiry. Keen holds that Watson "adamantly eschewed metaphysics altogether," and sees Watson's particular psychological approach as "a decision to create a science that was entirely separate from philosophy." He argues that "much of American psychology has followed this path, and psychology has continued to flourish as a science like physics, which has no inherent obligation to explore itself" (xii). Ludy Benjamin likewise shows how E. B. Titchener's method of introspection struggled against both functionalist psychology and Watsonian behaviorism. According to Benjamin, Titchener's psychological system died with him in 1927 as his students moved on to other projects and the new applied psychology increasingly pushed for an element of utility that Titchener had found at odds with his approach. Benjamin, *Brief History of Modern Psychology*. Kenton Kroker offers a slightly different history, suggesting that even as introspection lost favor as a scientific methodology in the

1920s and 1930s, it was taken up in a variety of therapeutic practices, including the "progressive relaxation" techniques of Edmund Jacobson. Kenton Kroker, "The Progress of Introspection in America, 1896–1938," *Studies in History and Philosophy of Biological and Biomedical Sciences* 34, no. 1 (2003): 77–108.

22. Wesley Mitchell, "Research in the Social Sciences," in *The New Social Science*, ed. Leonard Dupee White (Chicago: University of Chicago Press, 1930), 8.

23. Sigmund Freud, *Das Unbehagen in der Kultur* (Wien: Internationaler Psychoanalytischer Verlag, 1930).

24. Harold D. Lasswell to Charles Merriam, November 1, 1926, box 34, folder 4, CEMP. In a letter Lasswell wrote to his parents after hearing Bertrand Russell in Europe in 1923, he similarly stressed the centrality of physiological studies to social research. According to Lasswell, Russell "talked about science and the future of what we are pleased to call civilization, and what he said was perfectly brilliant. Of course he pointed out that science as such cannot save civilization, because science is merely a tool which may be used to kill or heal." As his own answer to the difficulties of science, Lasswell offered that "The problem is to work out a better technique for giving the kind and generous impulses the upper hand. This involves changes in social mechanism, in techniques of home training . . . and perhaps most hopefully on the progress of that group of students concerned with the physiology of the glands." Harold D. Lasswell to Anna and Linden Lasswell, October 31, 1923, box 56, folder 775, HDLP.

25. Harold D. Lasswell to Anna and Linden Lasswell, August 12, 1928, box 56, folder 782, HDLP.

26. Harold Lasswell, "Verbal References and Physiological Changes during the Psychoanalytic Interview: A Preliminary Communication," *Psychoanalytic Review* 22 (1935): 11. Lasswell reported additional technological research on emotion in Harold Lasswell, "Certain Prognostic Changes during Trial (Psychoanalytic) Interviews," *Psychoanalytic Review* 23, no. 3 (1936): 241–47.

27. Lasswell, "Verbal References and Physiological Changes," 12.

28. Camic, "On Edge" (see chap. 3, n. 23); Dorothy Ross, *The Origins of American Social Science* (Cambridge: Cambridge University Press, 1991); Bannister, *Sociology and Scientism* (see chap. 3, n. 23).

29. Camic, "On Edge," 231 (see chap. 3, n. 23).

30. Bannister, *Sociology and Scientism*, 3 (see chap. 3, n. 23).

31. Margaret Münsterberg, *Hugo Münsterberg: His Life and Work* (New York: Appleton, 1922); Merle J. Moskowitz, "Hugo Münsterberg: A Study in the History of Applied Psychology," *American Psychologist* 32, no. 10 (1977): 824–42; Giuliana Bruno, "Film, Aesthetics, Science: Hugo Münsterberg's Laboratory of Moving Images," *Grey Room*, no. 36 (2009): 88–113.

32. Hugo Münsterberg, *The Photoplay: A Psychological Study* (New York: Appleton, 1916), 112.

33. Ibid., 130.

34. Ibid., 126.

35. Ibid., 130. On Münsterberg's approach to film aesthetics, see Moskowitz, "Hugo Münsterberg"; Noël Carroll, "Film/Mind Analogies: The Case of Hugo Munsterberg," *Journal of Aesthetics and Art Criticism* 46, no. 4 (1988): 489–99; Vincent Colapietro, "Let's All Go to the Movies: Two Thumbs Up for Hugo Münsterberg's *The Photoplay* (1916)," *Transactions of the Charles S. Peirce Society* 36, no. 4 (2000): 477–501; and Bruno, "Film, Aesthetics, Science."

36. John R. Oliver, "Emotional States and Illegal Acts," *Journal of the American Institute of Criminal Law and Criminology* 11, no. 1 (1920): 84–85.

37. Joseph Roy Geiger, "The Effects of the Motion Picture on the Mind and Morals of the Young," *International Journal of Ethics* 34, no. 1 (1923): 69–70.

38. Ibid., 73.

39. Harmon B. Stephens, "The Relation of the Motion Picture to Changing Moral Standards," *Annals of the American Academy of Political and Social Science* 128 (1926): 151–57.

40. Peters, *Speaking into the Air* (see intro., n. 4); Martyn Jolly, *Faces of the Living Dead: The Belief in Spirit Photography* (West New York, NJ: Mark Batty, 2006); Tom Gunning, "To Scan a Ghost: The Ontology of Mediated Vision," *Grey Room*, no. 26 (2007): 94–127.

41. "Panic Riot in Theater," *Los Angeles Times*, August 27, 1911, I1; "Death Toll Needless," *Los Angeles Times*, August 28, 1911, I1; "Boys Caused Panic," *Washington Post*, August 28, 1911, 1.

42. Elizabeth Robins Pennell, "The Movies as Dope," *North American Review*, no. 792 (1921): 619.

43. R. Le Clero Phillips, "Modern Men's Emotions," *New York Times*, January 11, 1925, XX6.

44. For a discussion of the various voices participating in this censorship debate, see Robert Sklar, *Movie-Made America: A Social History of American Movies* (New York: Random House, 1975); and Czitrom, *Media and the American Mind*, esp. chap. 2 (see intro., n. 30). As Thomas Doherty explains, "though other media were more sexually explicit and politically incendiary, the domain of American cinema was panoramic and resonant, accessible to all, resisted by few. It was to Hollywood that politicians, clerics, and reformers looked when they detected a shredding of the moral fiber of the nation and a sickness in the body politic." Thomas Patrick Doherty, *Pre-Code Hollywood: Sex, Immorality, and Insurrection in American Cinema, 1930–1934* (New York: Columbia University Press, 1999), 319. On the relationship between the social sciences and film during this period, see also Mark Lynn Anderson, *Twilight of the Idols: Hollywood and the Human Sciences in 1920s America* (Berkeley: University of California Press, 2011).

45. "Advance Is Shown in Sound Recording," *New York Times*, December 10, 1931, 28.

46. Herman Scheffauer, "An Impression of the German Film *Metropolis*," *New York Times*, March 6, 1927, X7.

47. R. E. Sherwood, "*Metropolis*," *Life*, March 24, 1927, 24; Mordaunt Hall, "A Technical Marvel," *New York Times*, March 7, 1927, 16.

48. Margaret Walker to Harvie Clymer, September 30, 1926, container 1, folder 21, PFIR. For background on the Payne Fund and the Payne Fund studies, see Garth Jowett, I. C. Jarvie, and Kathryn H. Fuller, *Children and the Movies: Media Influence and the Payne Fund Controversy* (Cambridge: Cambridge University Press, 1996); McChesney, *Telecommunications, Mass Media, and Democracy* (see intro., n. 30); Czitrom, *Media and the American Mind* (see intro., n. 30); and Anderson, *Twilight of the Idols.*

49. Harvie Clymer to Benjamin Darrow, February 1, 1928, container 1, folder 21, PFIR; Mary K. Lakeman to Frances Payne Bolton, December 28, 1926, container 1, folder 22, PFIR.

50. Frances Payne Bolton to Lewis Lidyard, 1927, container 1, folder 23, PFIR.

51. Forman, *Our Movie Made Children*, 34–35.

52. Edgar Dale, *The Content of Motion Pictures* (New York: Macmillan, 1935), 225–26. The remaining Payne Fund monographs included Herbert Blumer, *Movies and Conduct* (New York: Macmillan, 1933); Herbert Blumer and Philip Morris Hauser, *Movies, Delinquency, and Crime* (New York: Macmillan, 1933); Werrett Wallace Charters, *Motion Pictures and Youth: A Summary* (New York: Macmillan, 1933); Edgar Dale, *How to Appreciate Motion Pictures*, experimental ed. (Columbus, OH, 1933); Dale, *Children's Attendance at Movies* (New York: Macmillan, 1935); Dysinger and Ruckmick, *Emotional Responses of Children*; Perry Ward Holaday and George Dinsmore Stoddard, *Getting Ideas from the Movies* (New York: Macmillan, 1933); Charles Clinton Peters, *Motion Pictures and Standards of Morality* (New York: Macmillan, 1933); Ruth Camilla Peterson and L. L. Thurstone, *Motion Pictures and the Social Attitudes of Children* (New York: Macmillan, 1933); Samuel Renshaw, Vernon Lemont Miller, and Dorothy Marquis, *Children's Sleep* (New York: Macmillan, 1933); Frank Kayley Shuttleworth and Mark Arthur May, *The Social Conduct and Attitudes of Movie Fans* (New York: Macmillan, 1933). The thirteenth manuscript, Paul Cressey's *Boys, Movies, and City Streets*, never made it into publication. See Jowett, Jarvie, and Fuller, *Children and the Movies*; and Jowett, Jarvie, and Fuller, "The Thirteenth Manuscript: The Case of the Missing Payne Fund Study," *Historical Journal of Film, Radio and Television* 13, no. 4 (1993): 387–402.

53. Charters, *Motion Pictures and Youth*; Holaday and Stoddard, *Getting Ideas from the Movies.*

54. Frances Payne Bolton, "Memorandum of Interview with Franklin Dunham," October 30, 1930, container 2, folder 25, PFIR.

55. "Biographical Sketch of Benjamin Darrow," 1927, container 3, folder 64, PFIR.

56. Benjamin Darrow, "Can the School Teach Discriminating Radio Listening?," 1932, container 55, folder 1054, PFIR.

57. J. L. Clifton, "Education by Radio," *Educational Research Bulletin* 9, no. 7 (1930): 199–201. On the larger history of the Ohio School of the Air, see T. C. Holy, "The Ohio School of the Air," *Educational Research Bulletin* 28, no. 6 (1949): 148–53; and McChesney, *Telecommunications, Mass Media, and Democracy* (see intro., n. 30).

58. "Gleanings from Various Sources," *Education by Radio* 5, no. 8 (1935): 30.

59. Carl E. Seashore, *Pioneering in Psychology* (Iowa City: University of Iowa Press, 1942), 11.
60. Stoddard, "Carl Emil Seashore" (see chap. 3, n. 56); Seashore, *Pioneering in Psychology*.
61. Carl E. Seashore, "The Growth of an Idea," *Scientific Monthly* 52, no. 5 (1941): 438.
62. Carl E. Seashore, "Some Psychological Statistics," *University of Iowa Studies in Psychology* 2 (1899): 1–84.
63. Seashore, "A Voice Tonoscope," 19 (see intro., n. 28).
64. Carl E. Seashore, ed., *University of Iowa Studies in Psychology*, vol. 6, The Psychological Monographs (Princeton, NJ: Psychological Review Company, 1914).
65. Seashore, "Tonoscope," 1 (see intro., n. 28).
66. Carl Seashore, "The Measure of a Singer," *Science* 35, no. 893 (1912): 204.
67. Seashore and Metfessel, "Deviation from the Regular," 538 (see chap. 3, n. 32).
68. Carl Seashore, "A Base for the Approach to Quantitative Studies in the Aesthetics of Music," *American Journal of Psychology* 39, no. 1 (1927): 141.
69. Seashore, "Phonophotography," 464 (see intro., n. 28).
70. Seashore, "Measure of a Singer," 211.
71. "Simon Guggenheim Gives $3,000,000 for Scholarships," *New York Times*, February 23, 1925, 1; "Dr. Dayton C. Miller Wins $1000 Prize of Science Association for Ether Drift Paper," *New York Times*, January 3, 1926, 1; "Yale 'Sound in Every Structure,' President Says in Commencement Talk," *New York Times*, June 20, 1935, 11; "Doctors Degrees Conferred on Quartet of Notables," *Los Angeles Times*, November 24, 1935, A3; "Pittsburgh Honors Two," *New York Times*, June 11, 1931, 23.
72. Rees Edgar Tulloss, "Address of Welcome," in *Feelings and Emotions: The Wittenberg Symposium*, ed. Martin L. Reymert (Worcester, MA: Clark University Press, 1928), 421.
73. At Seashore's request, the Payne Fund researchers sent Iowa a copy of an early report entitled "A Generation of Motion Pictures." "Addenda to Exhibits for Conference," November 24–25, 1928, container 29, folder 572, PFIR.
74. When Ruckmick decided to take the job at Iowa, he talked specifically about Seashore's work with his Cornell University mentor, E. B. Titchener. E. B. Titchener to Christian Ruckmick, March 13, 1924, box 5, PEBT.
75. Christian Ruckmich, "The Role of Kinaesthesis in the Perception of Rhythm," *American Journal of Psychology* 24, no. 3 (1913): 305–59. After graduate school, Ruckmick dropped the more Germanic spelling of "Ruckmich."
76. Christian Ruckmick, "An Institute for Acoustic Research," *Science* 56, no. 1448 (1922): 358.
77. Christian Ruckmick, "Why We Have Emotions," *Scientific Monthly* 28, no. 3 (1929): 255.
78. Ibid., 253.
79. Christian Ruckmick to William Short, September 26, 1930, container 28, folder 560, PFIR.
80. Ibid.

81. Christian Ruckmick to Walter B. Cannon, July 9, 1931, box 131, folder 1860, WBCP.

82. Christian Ruckmick, review of *The Fifth Column Is Here*, by George Britt, *Journal of Criminal Law and Criminology* 32, no. 1 (1941): 127.

83. E. B. Titchener, "Prolegomena to a Study of Introspection," *American Journal of Psychology* 23, no. 3 (1912): 433. On elements of Titchener's biography, see Howard C. Warren, "Edward Bradford Titchener," *Science* 66, no. 1705 (1927): 208–09; W. B. Pillsbury, "The Psychology of Edward Bradford Titchener," *Philosophical Review* 37, no. 2 (1928): 95–108; Donald A. Dewsbury, "Edward Bradford Titchener: Comparative Psychologist?," *American Journal of Psychology* 110, no. 3 (1997): 449–56; Benjamin, *Brief History of Modern Psychology*; and Edwin G. Boring, "Edward Bradford Titchener: 1867–1927," *American Journal of Psychology* 38, no. 4 (1927): 489–506. Ryan Tweney has complicated the common history of Titchener's work by pointing out that his version of introspection was a highly scientific, even positivist one. See Ryan D. Tweney, "Programmatic Research in Experimental Psychology: E. B. Titchener's Laboratory Investigations, 1891–1927," in *Psychology in Twentieth-Century Thought and Society*, ed. Mitchell G. Ash and William Ray Woodward (Cambridge: Cambridge University Press, 1987). On Titchener's method of introspection, see also Danziger, *Constructing the Subject*, 42–48; Kroker, "Progress of Introspection in America"; and Benjamin, *Brief History of Modern Psychology*.

84. Ruckmich, "Role of Kinaesthesis in the Perception of Rhythm."

85. Christian Ruckmick, "Experiences during Learning to Smoke," *American Journal of Psychology* 35, no. 3 (1924): 404.

86. Christian Ruckmich, "The History and Status of Psychology in the United States," *American Journal of Psychology* 23, no. 4 (1912): 531.

87. Christian Ruckmick, "On Overlooking Familiar Objects," *American Journal of Psychology* 37, no. 4 (1926): 632.

88. Ruckmich to Walter B. Cannon.

89. Ruckmich, "History and Status of Psychology in the United States."

90. Christian Ruckmick, "Empirical Psychology," *American Journal of Psychology* 40, no. 1 (1928): 167.

91. Christian Ruckmick, review of *A History of Experimental Psychology*, by Edwin G. Boring, *American Journal of Psychology* 42, no. 4 (1930): 657.

92. Dysinger and Ruckmick, *Emotional Responses of Children*, 18.

93. Ibid., 59.

94. As Kurt Danziger explains, "observer" (*Beobachter*) became one of the two standard terms for the subject of psychological experiments in the late nineteenth-century German research tradition developed by Wundt; "subject" (*Versuchsperson*) was the other. Danziger, *Constructing the Subject*, 32–33.

95. Christian Ruckmick, "How Do Motion Pictures Affect the Attitudes and Emotions of Children? The Galvanic Technique Applied to the Motion-Picture Situation," *Journal of Educational Psychology* 6, no. 4 (1932): 111.

96. Christian Cajavilca, Joseph Varon, and George L. Sternbach, "Luigi Galvani and the Foundations of Electrophysiology," *Resuscitation* 80, no. 2 (2009): 159–62; Marco Bresadola, "Animal Electricity at the End of the Eighteenth Century: The Many Facets of a Great Scientific Controversy," *Journal of the History of the Neurosciences* 17, no. 1 (2008): 8–32; Marco Piccolino, "The Bicentennial of the Voltaic Battery (1800–2000): The Artificial Electric Organ," *Trends in Neurosciences* 23, no. 4 (2000): 147–51; Piccolino, "Animal Electricity and the Birth of Electrophysiology: The Legacy of Luigi Galvani," *Brain Research Bulletin* 46, no. 5 (1998): 381–407; Marco Bresadola, "Medicine and Science in the Life of Luigi Galvani (1737–1798)," *Brain Research Bulletin* 46, no. 5 (1998): 367–80; Marco Piccolino, "Luigi Galvani and Animal Electricity: Two Centuries after the Foundation of Electrophysiology," *Trends in Neurosciences* 20, no. 10 (1997): 443–48; Marcello Pera, *The Ambiguous Frog: The Galvani-Volta Controversy on Animal Electricity* (Princeton: Princeton University Press, 1992); Sergio Trasatti, "1786–1986: Bicentennial of Luigi Galvani's Most Famous Experiments," *Journal of Electroanalytical Chemistry* 197, nos. 1–2 (1986): 1–4; P. Gallone, "Galvani's Frog: Harbinger of a New Era," *Electrochimica Acta* 31, no. 12 (1986): 1485–90.

97. Pera, *The Ambiguous Frog*, 19, 24.

98. Dr. Bischoff, "On Galvanism and Its Medical Applications," *Medical and Physical Journal* 7, no. 40 (1802): 529–40. On the history of the galvanometer, and the place of Bischoff within it, see R. S. Whipple, "The Evolution of the Galvanometer," *Journal of Scientific Instruments* 11, no. 2 (1920): 37–43.

99. Charles Féré, "Note sur les modifications de la tension électrique dans le corps humain," *Comptes rendus Société de biologie* 5 (1888): 28–33; J. Tarchanoff, "Über die galvanischen Erscheinungen in der Haut des Menschen bei Reizungen der Sinnesorgane und bei verschiedenen Formen der psychischen Thätigkeit," *Pflügers archiv für die gesammte physiologie* 46, no. 1 (1890): 46–55; Frederick Peterson and C. G. Jung, "Psycho-Physical Investigations with the Galvanometer and Pneumograph in Normal and Insane Individuals," *Brain* 30, no. 2 (1907): 153–218; E. W. Scripture, "Detection of the Emotions by the Galvanometer," *Journal of the American Medical Association* 50 (1908): 1164–65. On the history of psycho-galvanic measurements in psychology, see E. Prideaux, "The Psycho-galvanic Reflex: A Review," *Brain: A Journal of Neurology* 43, no. 1 (1920): 50–73; and Eva Neumann and Richard Blanton, "The Early History of Electrodermal Research," *Psychophysiology* 6, no. 4 (1970): 453–75.

100. Harold E. Jones and David Wechsler, "Galvanometric Technique in Studies of Association," *American Journal of Psychology* 40, no. 4 (1928): 607–12; David Wechsler, "On the Specificity of Emotional Reactions," *American Journal of Psychology* 36, no. 3 (1925): 424–26; David Wechsler and Harold E. Jones, "A Study of Emotional Specificity," *American Journal of Psychology* 40, no. 4 (1928): 600–606; Dysinger and Ruckmick, *Emotional Responses of Children*; Ruckmick, "How Do Motion Pictures Affect the Attitudes and Emotions of Children?," 210–19.

101. Christian A. Ruckmick and Emily Patterson, "A Simple Non-Polarizing Electrode," *American Journal of Psychology* 41, no. 1 (1929): 120–21; Christian Ruckmick, "A New Electrode for the Hathaway Galvanic Reflex Apparatus," *American Journal of Psychology* 42, no. 1 (1930): 106–7; Ruckmick, "Emotions in Terms of the Galvanometric Technique," *British Journal of Psychology* 21 (1930): 149–59; William H. Grubbs and Christian A. Ruckmick, "An Electrical Pneumograph," *American Journal of Psychology* 44, no. 1 (1932): 180–81.

102. Ruckmick, "Emotions in Terms of the Galvanometric Technique," 150.

103. Ruckmick, "How Do Motion Pictures Affect the Attitudes and Emotions of Children?," 210–11.

104. Dysinger and Ruckmick, *Emotional Responses of Children*, 1.

105. Ibid., 5–6.

106. Ibid., 16.

107. Larson joined Iowa's Psychopathic Hospital in the summer of 1930, and remained there for one year. August Vollmer to Dr. Guy Payne, December 19, 1929, box 1, folder 18, JALP; August Vollmer to John A. Larson, August 4, 1930, box 1, folder 18, JALP. While there, he had contact with the Psychology Department and with Carl Seashore. In a letter to August Vollmer after leaving Iowa, Larson explained, "If you remember I mentioned to you a conversation Dean Seashore, who is one of the country's leading psychologists, had with me. If I had remained in Iowa he offered to give me as many graduate students as I needed and as much apparatus, as he said I was the only one in a position on the firing line to keep the ill-trained expert from messing up the deception field." John Larson to August Vollmer, August 28, 1931, box 1, folder 18, JALP. Ruckmick had committed to work closely with Iowa's Psychopathic Hospital during his Payne Fund research. "The Influence of Motion Pictures on the Emotions of Children," Notes from a conference with Christian Ruckmick, October 28, 1929, container 28, folder 557, PFIR. William Short had likewise stressed the value of the research taking place in that part of the university and the importance of integrating its faculty into the Payne Fund studies. William Short to W. W. Charters, August 2, 1929, container 28, folder 557, PFIR. In 1938 Ruckmick wrote an essay entitled "The Truth about the Lie Detector" in which he drew on Larson's work. Impressed by Larson's research and the success of the lie detector in general, Ruckmick wrote that "with an admitted success of over 90 per cent accuracy in some ten thousand suspected cases, mostly in terms of a confession, however, the claims on behalf of this technique become impressive." Christian A. Ruckmick, "The Truth about the Lie Detector," *Journal of Applied Psychology* 22, no. 1 (1938): 51.

108. "Hands as Lie Detectors," *Washington Post*, January 9, 1938, TT13.

109. Darrell Huff, "Dead Pan No Defense with Emotion Meter," *Washington Post*, October 16, 1938, TS6.

110. Dysinger and Ruckmick, *Emotional Responses of Children*, 119.

111. Ibid.

112. Cartwright, "'Experiments of Destruction,'" 129.

113. Tom Gunning, "Renewing Old Technologies: Astonishment, Second Nature, and the Uncanny in Technology from the Previous Turn-of-the-Century," in *Rethinking Media Change: The Aesthetics of Transition*, ed. David Thorburn, Henry Jenkins, and Brad Seawell (Cambridge: MIT Press, 2003), 39–60; Gunning, "To Scan a Ghost."

114. Christian A. Ruckmick, "Discussion: Terminology in Re 'Psychogalvanic Reflex,'" *Psychological Review* 40, no. 1 (1933): 97–98; Ruckmick, "A Critique of the 'Galvanic' Technique," *Psychological Review* 45, no. 2 (1938): 154–62.

115. McChesney, *Telecommunications, Mass Media, and Democracy* (see intro., n. 30); McChesney, *Rich Media, Poor Democracy* (see intro., n. 30).

116. Grace Kingsley, "Universal Signs Psychologist: Inventor of Lie Detector Will Analyze U. Pictures," *Los Angeles Times*, January 11, 1929, A10; Philip Scheuer, "Film Kiss Holds Dynamite: Noted Psychologist Employed by Large Picture Producing Concern Tells of Problems," *Los Angeles Times*, July 14, 1929, B13.

117. Titchener to Ruckmick, March 13, 1924.

118. According to a news article in the *Des Moines Register*, "the university administration didn't approve of the thing at all—or in the technique Dr. Ruckmick used in employing the instrument with co-eds and men students of the university as subjects." It continues, "the subjects—students taking [Ruckmick's] required course—had their emotion reactions measured in steel-lined, soundproof, locked rooms, with nobody present but the student and instructor." "S.U.I Faculty Row Laid to Emotion Test," *Des Moines Register*, January 29, 1938. The correspondence between Ruckmick, Dean George Kay, university president Eugene Gilmore, and others suggests a much more complicated picture, including regarding whether or not Ruckmick had actually resigned his position. Apparently, Ruckmick suggested that he would be forced to resign if certain conditions were not met—primarily regarding who chaired the department of psychology—and the dean and president accepted this as his official resignation. Ruckmick's attempts to deny or rescind this resignation were refused by the university administration, and he left the university during the 1938 school year. "Christian Ruckmick," Faculty and Staff Vertical Files (RG 01.15.03), UAUI.

119. Ruckmick served as a secretary and general sales manager for C. H. Stoelting Company from 1938 to 1942. From 1942 to 1943, he was employed as a psychologist for the U.S. Air Force. In the late 1940s and into the 1950s, Ruckmick worked as the minister of education and fine arts for Ethiopia. *Amherst College Biographical Record, 1963: Biographical Record of the Graduates and Non-Graduates of the Classes of 1822–1962 Inclusive* (Amherst, MA: Trustees of Amerst College, 1963), 183; Christian Ruckmick to Norman Charles Meier, box 1, Acknowledgments from vol. 3, Papers of Norman Meier (RG 99.0163), UAUI.

120. Seashore, *Manual of Instructions* (see chap. 3, n. 55); Ruckmick, "How Do Motion Pictures Affect the Attitudes and Emotions of Children?"

121. The Papers of Carl Seashore (RG 99.0164), UAUI, contain a number of references to Seashore's work with Columbia and Victor; Oversize Box 1 contains

sets of records by both Columbia and RCA Victor. Box 2 contains a draft of a contract between Seashore and the Columbia Graphophone Company. Seashore would be paid 15 percent of 75 percent of the value of records in which both sides contained songs created solely by him, and 7.5 percent of 75 percent of the value of records in which only one side contained songs created solely by him.

122. "Invention Records All Heart Sounds," *New York Times*, November 1, 1925, 27.

123. "New Instrument in Medicine Built Like Radio Receiving Set: Electro-Cardiograph Developed by the General Electric Company Renders Heart 'Voltage' Visible," *New York Times*, December 14, 1924, XX18.

124. Several critiques of empirical methods of media research have demonstrated how these quantitative studies tend to reinforce commercial imperatives by framing the relationship between media and audiences as an individual rather than more broadly social phenomenon. By taking individual audience members as a primary focus (even in aggregate form), this research generally ignores cultural and economic factors important to how media messages are produced and consumed. Instead, the media-audience relationship takes the form most desirable to commercial broadcasters by considering how individual consumers do or do not respond to a particular message. See Carey, *Communication as Culture* (see intro., n. 30); Hanno Hardt, *Critical Communication Studies: Communication, History, and Theory in America* (London: Routledge, 1992); and Todd Gitlin, "Media Sociology: The Dominant Paradigm," *Theory and Society* 6, no. 2 (1978): 205–53. Paul Lazarsfeld's classic discussion of "administrative" and "critical" research admits to this distinction as well, seeing some research approaches as more appropriate for the practical decision making of commercial broadcasters and government agencies. Lazarsfeld, "Remarks on Administrative and Critical Research" (see intro., n. 29).

125. William Short to Harvie Clymer, November 15, 1927, container 1, folder 23, PFIR.

126. Mark Smith's *Social Science in the Crucible* does an especially strong job of showing how funding issues tended to overdetermine the kinds of social research being performed in the 1920s and 1930s. Because of their orientation toward public health and education, the funding agencies of this period tended to support research with a clear angle on the diagnosis or prevention of public problems. This pushed media researchers toward questions of media effects or toward the educational benefits of mass media—another research area of the Payne Fund. These areas also tended to be the most politically safe, allowing funding agencies to avoid entangling themselves in more complicated political or economic struggles. Smith, *Social Science in the Crucible* (see chap. 3, n. 23).

127. Charters, *Motion Pictures and Youth*, 29–31.

128. Blumer, *Movies and Conduct*, 300.

129. "FRC Interpretation of Public Interest," in *Documents of American Broadcasting*, ed. Frank J. Kahn (Englewood Cliffs, NJ: Prentice Hall, 1984), 57–62.

130. McChesney, *Telecommunications, Mass Media, and Democracy* (see intro., n. 30).

131. J. H. Dellinger, "Analysis of Broadcasting Station Allocation," *Proceedings of the Institute of Radio Engineers* 16, no. 11 (1928): 1477–85.

132. Nicoline Grinager Ambrose and Ehud Yairi, "The Tudor Study: Data and Ethics," *American Journal of Speech-Language Pathology* 11, no. 2 (2002): 190–203; Mary Tudor, "An Experimental Study of the Effect of Evaluative Labeling on Speech Fluency" (unpublished MA thesis, University of Iowa, 1939).

133. Dysinger and Ruckmick, *Emotional Responses of Children*, 17.

134. Ruckmick, "How Do Motion Pictures Affect the Attitudes and Emotions of Children?," 214. According to Jon Solomon, Walsh's *Wanderer* was "probably the most financially successful silent film about Babylon." Jon Solomon, *The Ancient World in the Cinema* (New Haven: Yale University Press, 2001), 227.

135. Dysinger and Ruckmick, *Emotional Responses of Children*, 80.

136. Ibid., 82.

137. Ibid., 73.

138. "Child's Reactions to Movies Is Shown," *New York Times*, May 28, 1933, E7.

139. Harold D. Lasswell, "The Structure and Function of Communication in Society," in *The Communication of Ideas: A Series of Addresses*, ed. Lyman Bryson (New York: Institute for Religious and Social Studies, 1948), 37–51. On the origin of Lasswell's chart, see William Buxton, "From Radio Research to Communications Intelligence: Rockefeller Philanthropy, Communications Specialists, and the American Policy Community," in *The Political Influence of Ideas: Policy Communities and the Social Sciences*, ed. Stephen Brooks and Alain Gagnon (Westport, CT: Praeger, 1994), 187–209. Buxton suggests that this chart originated in the same philanthropic climate that produced the Payne Fund motion picture studies.

140. Carey, *Communication as Culture* (see intro., n. 30).

141. Communications Act of 1934, Public Law 416, *U.S. Statutes at Large* 48 (1934): 1064 (emphasis added).

NOTES TO CHAPTER 5

1. Matt Richtel, "Digital Devices Deprive Brain of Needed Downtime," *New York Times*, April 24, 2010.

2. Mattias Karlsson and Loren Frank, "Awake Replay of Remote Experiences in the Hippocampus," *Nature Neuroscience* 12 (2009): 913.

3. Howard Taubman, "Creativity and Computers: A View That Electronic Systems Will Liberate Artists from Routine," *New York Times*, June 14, 1966, 52; Otto Friedrich, "Machine of the Year 1982: The Computer Moves In," *Time*, January 3, 1983, 14; Henry Jenkins, *Convergence Culture: Where Old and New Media Collide* (New York: New York University Press, 2006); Mark Poster, *Information Subject* (New York: Routledge, 2001); Al Gore, "Bringing Information to the World: The Global Information Infrastructure Policy Commentaries," *Harvard Journal of Law and Technology* 9 (1996). Celebrating the work of Alvin and Heidi Toffler, Newt Gingrich told a story of liberation via digital technology: "On January 5, 1995, the Third Wave came to American Democracy in the form of 'Thomas,' the Library of Congress's on-line system that allows every citizen to access copies of legislation, committee reports and other congressional

documents. During its first four days of operation, 28,000 individuals and 2,500 institutions used Thomas to download 175,132 documents." Newt Gingrich, foreword to *Creating a New Civilization: The Politics of the Third Wave*, by Alvin Toffler and Heidi Toffler (Atlanta: Turner, 1995), 17. John P. Foley, "The Church and Internet," *Pontifical Council for Social Communications* (2002), http://www.vatican.va/roman_curia/pontifical_councils/pccs/documents/rc_pc_pccs_doc_20020228_church-internet_en.html.

4. Friedrich, "Machine of the Year 1982"; Postman, *Technopoly*, 107 (see chap. 2, n. 31); Neil Postman, *Amusing Ourselves to Death: Public Discourse in the Age of Show Business* (New York: Penguin, 1986); Douglas Birch, "Just a Little Too Tangled Up in the Internet," *Los Angeles Times*, September 5, 1994, 3.

5. Bob Strauss, "Mind over *Matrix*," *Daily News of Los Angeles*, March 31, 1999.

6. In an early scene of the film, Neo pulls a computer disk from a hollowed-out copy of Baudrillard's book *Simulacra and Simulation*. Jean Baudrillard, *Simulacra and Simulation* (Ann Arbor: University of Michigan Press, 1994).

7. Birch, "Just a Little Too Tangled Up," 3.

8. Sharon Begley, "I Can't Think! The Twitterization of Our Culture Has Revolutionized Our Lives, but with an Unintended Consequence—Our Overloaded Brains Freeze When We Have to Make Decisions," *Newsweek*, March 7, 2011, 28.

9. "The Teenage Brain in the Digital Age," *NBC Nightly News*, April 12, 2011.

10. Steven Johnson, *Everything Bad Is Good for You: How Today's Popular Culture Is Actually Making Us Smarter* (New York: Riverhead, 2006).

11. Ibid., 155.

12. Ibid., 33.

13. Ibid.

14. Ibid., 35.

15. Ibid., 34.

16. Nicholas G. Carr, *The Shallows: What the Internet Is Doing to Our Brains* (New York: Norton, 2011).

17. Ibid., 7.

18. Nicholas G. Carr, "Is Google Making Us Stupid? What the Internet Is Doing to Our Brains," *Atlantic*, July/August 2008.

19. Carr, *The Shallows*, 170.

20. Ibid., 3 (emphasis added).

21. Ibid., 20.

22. Ibid., 71–72.

23. Steven Johnson, *Mind Wide Open: Your Brain and the Neuroscience of Everyday Life* (New York: Scribner, 2004).

24. Ibid., 2.

25. Ibid., 3.

26. Johnson, *Everything Bad Is Good for You*, 155–56.

27. Carr, *The Shallows*, 74.

28. Eleanor A. Maguire et al., "Navigation-Related Structural Change in the Hippocampi of Taxi Drivers," *Proceedings of the National Academy of Sciences* 97, no. 8 (2000): 4398.

29. Johnson, *Everything Bad Is Good for You*, 108.

30. Carr, *The Shallows*, 32–33.

31. Daniel R. Anderson et al., "Brain Imaging: An Introduction to a New Approach to Studying Media Processes and Effects," *Media Psychology* 8, no. 1 (2006): 6.

32. Gary W. Small and Gigi Vorgan, *iBrain: Surviving the Technological Alteration of the Modern Mind* (New York: Collins Living, 2008).

33. Ibid., 73–74.

34. Ibid., 80–81.

35. Ibid., 88.

36. Ibid., 12. Gutenberg's printing press was developed sometime in the fifteenth century, and Morse put electricity—previously discovered—to work for the telegraph in the 1840s. The telephone was patented in the late nineteenth century, but not fully utilized until the early twentieth.

37. Kelly A. Joyce, *Magnetic Appeal: MRI and the Myth of Transparency* (Ithaca: Cornell University Press, 2008), 11.

38. William R. Uttal, *The New Phrenology: The Limits of Localizing Cognitive Processes in the Brain* (Cambridge: MIT Press, 2001), xi.

39. Ibid., 16.

40. Carr borrows this phrase from David Golumbia, *The Cultural Logic of Computation* (Cambridge: Harvard University Press, 2009). However, Carr's is a very selective reading of Golumbia's book in that it ignores Golumbia's discussion of the functionalism of Jerry Fodor. When Fodor subscribed to a reductive understanding of the mind—the sort that underlies the physicalist enterprise—Golumbia suggests that he imposed the cultural logic of computation onto the processes of cognition. Significantly, the philosopher Daniel Dennett, who is a longtime advocate of using neurophysiological data to understand human experience, is also heavily associated with a "computational turn" in philosophical thinking about the mind. Jon Dorbolo, "Daniel Dennett and the Computational Turn," *Minds and Machines* 16, no. 1 (2006): 1.

41. Johnson, *Everything Bad Is Good for You*, 142.

42. On the history of people's various uses of the book, see Ronald J. Zboray and Mary Saracino Zboray, *Literary Dollars and Social Sense: A People's History of the Mass Market Book* (New York: Routledge, 2005); and Zboray and Zboray, *Everyday Ideas: Socioliterary Experience among Antebellum New Englanders* (Knoxville: University of Tennessee Press, 2006). Thomas Jefferson, *The Jefferson Bible: The Life and Morals of Jesus of Nazareth Extracted Textually from the Gospels in Greek, Latin, French and English*, Smithsonian ed. (Washington, DC: Smithsonian Books, 2011). The narrow, ahistorical conception of "reading" that Carr evokes is also problematic for Maryanne Wolf, from whom Carr draws much of

his argument. Maryanne Wolf, *Proust and the Squid: The Story and Science of the Reading Brain* (New York: HarperCollins, 2007).

43. Carr, *The Shallows*, 217.

44. Small and Vorgan, *iBrain*, 9.

45. Ibid., 91.

46. Rosalind W. Picard, *Affective Computing* (Cambridge: MIT Press, 1997).

47. Ibid., 104.

48. The website is www.affectiva.com.

49. On facial recognition software, see Kelly Gates, *Our Biometric Future: Facial Recognition Technology and the Culture of Surveillance* (New York: New York University Press, 2011).

50. Jaak Panksepp, "Oxytocin Effects on Emotional Processes: Separation Distress, Social Bonding, and Relationships to Psychiatric Disorders," *Annals of the New York Academy of Sciences* 652 (1992): 243–53; Panksepp, "The Emotional Sources of 'Chills' Induced by Music," *Music Perception: An Interdisciplinary Journal* 13, no. 2 (1995): 171–207; Jaak Panksepp and Günther Bernatzky, "Emotional Sounds and the Brain: The Neuro-Affective Foundations of Musical Appreciation," *Behavioural Processes* 60, no. 2 (2002): 133–55; personal communication from Jaak Panksepp to Brenton J. Malin, August 22, 2011.

51. Thomas Nagel, "What Is It Like to Be a Bat?," *Philosophical Review* 83, no. 4 (1974): 435–50.

52. Thomas Nagel, "Physicalism," *Philosophical Review* 74, no. 3 (1965): 339.

53. Nagel, "What Is It Like to Be a Bat?," 435; Thomas Nagel, review of *Content and Consciousness*, by Daniel Dennett, *Journal of Philosophy* 69, no. 8 (1972): 220–24.

54. Daniel C. Dennett, *Content and Consciousness* (New York: Humanities Press and Routledge and Kegan Paul, 1969), ix.

55. Ibid., 88.

56. Nagel, "What Is It Like to Be a Bat?," 436.

57. Ibid., 438.

58. Ibid., 439.

59. Ibid., 443.

60. Frank Jackson, "Epiphenomenal Qualia," *Philosophical Quarterly* 32, no. 127 (1982): 127–36. Jackson expands this argument in "What Mary Didn't Know," *Journal of Philosophy* 83, no. 5 (1986): 291–95.

61. Jackson, "Epiphenomenal Qualia," 130.

62. Ibid., 127.

63. Paul M. Churchland, "Subjective Qualia from a Materialist Point of View," *PSA: Proceedings of the Biennial Meeting of the Philosophy of Science Association*, 1984, 773–90; Churchland, "Reduction, Qualia, and the Direct Introspection of Brain States," *Journal of Philosophy* 82, no. 1 (1985): 8–28; David J. Chalmers and Frank Jackson, "Conceptual Analysis and Reductive Explanation," *Philosophical Review* 110, no. 3 (2001): 315–60; Peter Ludlow, Yujin Nagasawa, and Daniel Stoljar, eds., *There's Something about Mary: Essays on Phenomenal Consciousness and Frank*

Jackson's Knowledge Argument (Cambridge: MIT Press, 2004); Craig Delancey, "Phenomenal Experience and the Measure of Information," *Erkenntnis (1975–)* 66, no. 3 (2007): 329–52; Torin Andrew Alter and Sven Walter, *Phenomenal Concepts and Phenomenal Knowledge: New Essays on Consciousness and Physicalism* (Oxford: Oxford University Press, 2007); David Chalmers, *The Conscious Mind: In Search of a Fundamental Theory* (New York: Oxford University Press, 1996).

64. Daniel C. Dennett, "What RoboMary Knows," in Alter and Walter, *Phenomenal Concepts and Phenomenal Knowledge*, 15–31. Dennett introduced an earlier version of this thought experiment in *Sweet Dreams: Philosophical Obstacles to a Science of Consciousness* (Cambridge: MIT Press, 2005). He offered a still earlier critique of Mary's room in *Consciousness Explained* (Boston: Little, Brown, 1991).

65. Churchland, "Reduction, Qualia, and the Direct Introspection of Brain States," 25–26 (emphasis added).

66. Chalmers, *The Conscious Mind*, 141.

67. David Chalmers, "Facing up to the Problem of Consciousness," *Journal of Consciousness Studies* 2, no. 3 (1995): 200. For a further refutation of Dennett's RoboMary discussion, see Michael Beaton, "What RoboDennett Still Doesn't Know," *Journal of Consciousness Studies* 12, no. 12 (2005): 3–25.

68. Dennett, "What RoboMary Knows," 21.

69. Helen Keller, *The Story of My Life*, ed. John Albert Macy (Garden City, NY: Doubleday, Page, 1921), 253.

70. Marshall McLuhan, *The Gutenberg Galaxy: The Making of Typographic Man* (Toronto: University of Toronto Press, 1962), 37; John Wilson, "Film Literacy in Africa," *Canadian Communications* 1 (1961): 7–15.

71. Wilson, "Film Literacy in Africa," 9.

72. McLuhan, *Gutenberg Galaxy*, 37.

73. James Burns, "Watching Africans Watch Films: Theories of Spectatorship in British Colonial Africa," *Historical Journal of Film, Radio and Television* 20, no. 2 (2000): 197–211.

74. Ibid.

75. P. Morton-Williams, *Cinema in Rural Nigeria: A Field Study of the Impact of Fundamental-Education Films on Rural Audiences in Nigeria* (Ibadan: Federal Information Service, 1953), 44; Burns, "Watching Africans Watch Films," 206.

76. Renée Hobbs et al., "How First-Time Viewers Comprehend Editing Conventions," *Journal of Communication* 38, no. 4 (1988): 54.

77. Jenna Burrell, *Invisible Users: Youth in the Internet Cafés of Urban Ghana* (Cambridge: MIT Press, 2012).

78. Ibid., 55.

79. Ibid., 59.

80. Burns, "Watching Africans Watch Films," 198.

81. Naomi Scheman, "Individualism and the Objects of Psychology," in *Engenderings: Constructions of Knowledge, Authority, and Privilege* (New York: Routledge, 1993), 37.

82. Donna Jeanne Haraway, *Modest_Witness@Second_Millennium.Femaleman©_Meets_ Oncomouse™: Feminism and Technoscience* (New York: Routledge, 1997), 142–44.

83. Michael Warner, "The Mass Subject and the Mass Public," in *Habermas and the Public Sphere*, ed. Craig J. Calhoun (Cambridge: MIT Press, 1994), 377–401; Warner, *Publics and Counterpublics* (New York: Zone, 2002).

84. Antonio R. Damasio, *Descartes' Error: Emotion, Reason, and the Human Brain* (London: Penguin, 2005); Damasio, *The Feeling of What Happens: Body and Emotion in the Making of Consciousness* (New York: Harcourt Brace, 1999).

85. Damasio, *Descartes' Error*, 208.

86. Ibid., 209.

87. Antonio R. Damasio, Daniel Tranel, and Hannah Damasio, "Somatic Markers and the Guidance of Behavior: Theory and Preliminary Testing," in *Frontal Lobe Function and Dysfunction*, ed. Harvey S. Levin, Howard M. Eisenberg, and Arthur Lester Benton (New York: Oxford University Press, 1991), 222.

88. If we don't take for granted that a disturbing emotional image can be established without some reference to a broader culture, then we might well assume that Damasio's patients are responding to the social context of emotion rather than to a given feeling evoked—or not—by a specific image. From this perspective, it could be just as likely that the frontal lobe helps people make connections to the broader culture, rather than allowing them to feel emotions per se. In any case, Damasio's claims suggest the problems of attempting to localize cultural ideals—such as disturbingness—to specific regions of the brain.

89. Daniel Gross offers a similar critique of a study in which Damasio asked people with amygdala damage to identify the trustworthiness of various people's faces. Here again, Damasio ignores the larger social and cultural context in which a notion of trustworthiness would make sense. Gross, *Secret History of Emotion*, 30–31 (see intro., n. 33).

90. Kenneth Burke, *Language as Symbolic Action: Essays on Life, Literature, and Method* (Berkeley: University of California Press, 1966), 49.

91. Harold Innis, "The Newspaper in Economic Development," *Journal of Economic History* 2 (1942): 1–33; Innis, "On the Economic Significance of Culture," *Journal of Economic History* 4 (1944): 80–97; Innis, *Empire and Communications* (Oxford: Clarendon, 1950); Innis, *The Bias of Communication* (Toronto: University of Toronto Press, 1951).

92. Marshall McLuhan, *Understanding Media: The Extensions of Man* (1964; repr., Cambridge: MIT Press, 1994), 22–32; McLuhan, *Gutenberg Galaxy*, 19, 138.

93. Marshall McLuhan and Eric McLuhan, *Laws of Media: The New Science* (Toronto: University of Toronto Press, 1988), 69.

94. Claude Elwood Shannon and Warren Weaver, *The Mathematical Theory of Communication* (Urbana: University of Illinois Press, 1949).

95. McLuhan and McLuhan, *Laws of Media*, 86.

96. Eric Schwitzgebel, *Perplexities of Consciousness* (Cambridge: MIT Press, 2011), 15.

97. Ibid., 123.

98. Jerome S. Bruner, *Acts of Meaning* (Cambridge: Harvard University Press, 1990), 14.

99. Telecommunications Act of 1996, section 3 (48).

100. McChesney, *Rich Media, Poor Democracy* (see intro., n. 30).

101. Jim Exon, "The Communications Decency Act," *Federal Communications Law Journal* 49, no. 1 (1996): 95.

NOTES TO THE CONCLUSION

1. *Jeopardy!*, February 14, 2011 (emphasis added).

2. John Markoff, "Computer Wins on *Jeopardy!*: Trivial, It's Not," *New York Times*, February 17, 2011, A1.

3. John Markoff, "A Fight to Win the Future: Computers vs. Humans," *New York Times*, February 15, 2011, D1.

4. John R. Searle, "Watson Doesn't Know It Won on *Jeopardy!*," *Wall Street Journal*, February 23, 2011. In Searle's "Chinese room" argument, he asks people to imagine a person inside a small room. The person inside the room does not read or speak Chinese. Questions written using Chinese characters are passed through a hole in the wall of the room. Using an elaborate instruction book—written in his or her native language—the person inside puts together some Chinese characters and then passes them through another hole in the wall. As far as someone who understands Chinese is concerned, the person in the box has answered the question. The person inside the box, of course, has no idea either what he or she was asked or what he or she answered. Searle's point is that manipulating symbols—what computers do—is not the same as understanding meaning. John R. Searle, "Minds, Brains, and Programs," *Behavioral and Brain Sciences* 3 (1980): 417–57. See also Searle, "Is the Brain a Digital Computer?," *Proceedings and Addresses of the American Philosophical Association* 64, no. 3 (1990).

5. *Jeopardy!*, February 15, 2011.

6. Jim Giles, "What's Next for Watson?," *New Scientist*, February 19, 2011.

7. Lenore Skenazy, "Watson Victory on *Jeopardy* Purely Elementary," *San Gabriel Valley Tribune*, February 19, 2011.

8. Bob Fredericks, "*Jeopardy!* Champ Now Dr. Watson," *New York Post*, February 18, 2011, 9.

9. Ken Jennings, "My Puny Human Brain," *Slate*, February 17, 2011.

10. Joshua Gunn, "On Speech and Public Release," *Rhetoric and Public Affairs* 13, no. 2 (2010): 184.

11. On the history of Western medicine, especially in the American context, as well as its relationship to "alternative medicine," see, for instance, Paul Starr, *The Social Transformation of American Medicine: The Rise of a Sovereign Profession and the Making of a Vast Industry* (New York: Basic, 1982); W. F. Bynum, *The Western Medical Tradition: 1800 to 2000* (New York: Cambridge University Press, 2006); and Roberta E. Bivins, *Alternative Medicine? A History* (Oxford: Oxford University Press, 2007).

12. Peter Simonson and Gabriel Weimann, "Critical Research at Columbia: Lazars-
feld and Merton's 'Mass Communication, Popular Taste, and Organized Social
Action,'" in *Canonic Texts in Media Research: Are There Any? Should There Be?
How about These?*, ed. Elihu Katz, John Durham Peters, Tamar Liebes, and Avril
Orloff (Cambridge: Polity, 2003), 12–38.

13. Richard J. Franke, "The Power of the Humanities and a Challenge to Human-
ists," *Daedalus* 138, no. 1 (2009): 13.

14. The work of the art historian David Freedberg offers a strong example of technologically centered research on art. See David Freedberg and Vittorio Gallese,
"Motion, Emotion and Empathy in Esthetic Experience," *Trends in Cognitive Sciences* 11, no. 5 (2007): 197–203; Fortunato Battaglia, Sarah H. Lisanby, and David
Freedberg, "Corticomotor Excitability during Observation and Imagination of
a Work of Art," *Frontiers in Human Neuroscience* 5 (2011): 1–6; Vittorio Gallese
and David Freedberg, "Mirror and Canonical Neurons Are Crucial Elements in
Esthetic Response," *Trends in Cognitive Sciences* 11, no. 10 (2007): 411. For a summary of recent scientific research on literature, and a critique of it, see Jonathan
Kramnick, "Against Literary Darwinism," *Critical Inquiry* 37, no. 2 (2011): 315–47.

15. Damasio, *The Feeling of What Happens* (see chap. 5, n. 84).

16. Kittler, *Discourse Networks, 1800/1900*, 369 (see intro., n. 30).

17. Ibid., 370.

ABOUT THE AUTHOR

Brenton J. Malin is Associate Professor in the Department of Communication and an affiliate faculty member in the Cultural Studies Program at the University of Pittsburgh.